The Hand and the Brain

Göran Lundborg

The Hand and the Brain

From Lucy's Thumb
to the Thought-Controlled Robotic Hand

 Springer

Göran Lundborg, MD, PhD
Hand Surgery
Department of Clinical Sciences
Malmö Lund University
Skäne University Hospital
Malmö
Sweden

ISBN 978-1-4471-5333-7   ISBN 978-1-4471-5334-4   (eBook)
DOI 10.1007/978-1-4471-5334-4
Springer London Heidelberg New York Dordrecht

Library of Congress Control Number: 2013949173

Printed on acid-free paper

Springer is part of Springer Science+Business Media (www.springer.com)

To my family

Preface

The hand has been called an extension of the brain, an outer brain and a mirror of the soul. Much of our personality and identity is contained in the gestures and movements of the hands. The hand reflects our mind, expressing our innermost thoughts and wishes. With our hands we communicate with people. The hands plead, bless and love; they express hope, despair, disgust and hatred. They welcome, caress and punish.

In my career as a hand surgeon, I have seen the most injured, painful and malformed hands, and I have authored books, including several textbooks explaining how to provide patients with the best treatments. But why shouldn't healthy, noninjured hands also be worthy of a book? All their amazing experience and tacit accumulated knowledge offer an enormous wealth of possibilities. Thanks to our hands we can build and operate computers, key in correct PIN codes and use the TV remote control. Creative hands are the basis of our civilization; they construct monuments, build bridges, create parks and build skyscrapers; they craft symphonies, make musical instruments, bring about pleasant music and create great works of art.

But the hand is not only the brain's delicate and fine-tuned instrument – it is also a sense organ. The sense of touch and the hyper-advanced sensory functions of the hand make it possible to examine and explore our surrounding world. Our sense of touch lets us discover the true character of what the eye sees – *seeing is believing, but touching is understanding*. The hand 'sees' in dark, and the sense of touch can even substitute for the loss of other senses; a blind person reads the Braille alphabet with his fingertips. The sense of touch makes it possible to differentiate silk from velvet, and the fingertips easily discover scratches and stray grains of salt on the surface of a table. Representation of the hand takes up a very large area in the brain, and this area expands as sensory inflow and hand activity increase – yes, the hand shapes the brain. Perhaps we can even regard the brain as an extension of the hand into our soul.

If a hand is missing from birth due to a congenital anomaly or an accident, is there any way to substitute for a lost hand? Can a hand be transplanted from one individual to another? Or can an artificial hand substitute for an actual hand and intuitively be controlled by the mind and capable of executing the same movements and grip functions as an actual hand?

And how did the story begin? Today we know that the hand was highly advanced more than 4 million years ago, at a time when our ancestors' brains were still rudimentary and the size of a chimpanzee's. Perhaps it was the advanced, agile hand, with all its promise for tool making, gestures and signs, that stimulated the brain's development. Perhaps we are indebted to the early hand for the development of our cognitive capacity and intelligence as well as our capacities for abstract thinking, language, art and music.

So the saga of the hand is fascinating. As you read this book, you will gain insights you have never thought of before, and you will be surprised when you realise the role of hands in our culture and in our life. So sit back in your favourite chair, relax, read and get to know yourself – or, more precisely, your hand.

Malmö, Sweden   Göran Lundborg, MD, PhD

Acknowledgements

I am very grateful to friends, co-workers and experts who have contributed in various ways to the birth of this book, especially Christian Antfolk, Bo Balldin, Lasse Bengtsson, Anders Björkman, Isabella Björkman-Burtscher, Ingela Brovik, Rafael Cierpka, Christian Cipriano, Marco Controzzi, Jan Delden, Robin af Ekenstam, Henrik Ehrsson, Hans-Peter Feldmann, Thomas Hansson, Björn Henriksson, Anna Holmbom, Mats Huddén, Gunnar Jansson, Anders Jarlert, Marco Lanzetta, Bengt Nilsson, Gabriela Pichler, Birgitta Rosén, Fredrik Sebelius, Niklas Schiöler, Viola Sörensson, Jimmy Wahlstedt, Andreas Weibull and Bertil Widenfalk.

I am very grateful to Professor Dean Snow of Pennsylvania State University, who very generously allowed me to use the handprints from the Gargas and Pech Merle caves that he personally documented.

A special thanks to my artist Fredrik Johansson and my secretary Tina Folker, who have been an enormous help in designing and processing the text and references.

I also want to express my great appreciation to Linda Evans and Jennifer Evans of Changeling Translations AB, who corrected and improved the quality of my English text in a very skilled and professional way.

And finally, a warm thanks to my dear wife Christina, who has been unfailing in her support and a constant source of inspiration.

Generous financial support has been provided by the Evy and Gunnar Sandberg Foundation and the Royal Physiographic Society in Lund.

Contents

Chapter 1
Fins, Fossils and Fingers

Abstract Three hundred and seventy-five million years ago, many different types of big fish lived in the sea. Gradually, some of them began approaching shallow water close to land. It became advantageous for these fishes to be able to support themselves on the seabed with their fins so they could raise their heads above the water's surface in an amphibian-like way. In this transitional phase between life in the water and life on land, the fins of some fish species showed an obvious development towards an arm and a hand. The *Tiktaalik*, discovered on Ellesmere Island in northern Canada and dating back to about 375 million years ago, has been regarded as a missing link between fish and land animals, showing a first hint of a human hand in its fin.

We tend to take our hands for granted, always present and always prepared to intuitively translate our wishes and intentions into movement – whether turning pages in a newspaper, peeling a freshly boiled egg, shaking hands with a good friend or unlocking the front door when we return home from work. The hand is a fine motor instrument with great precision, but it is also a delicate organ with highly advanced sensory functions.

When in the evolutionary process can we first distinguish a preliminary draft of something that would much later result in a human hand? To find an answer, we have to go back several hundred million years, to a period of the earth's development when no land animals yet existed.

About 375 million years ago when the earth was mainly covered with water, the remaining land was swampy, marshy, and filled with mangrove-like trees. Many different species of big fish lived in the sea, some of them with swim bladders that would later develop into lungs [1]. Thanks to rich vegetation, with plants and trees growing close to the seashores, the water became increasingly nutrient-rich, and some fish species began approaching the shallow water close to the mainland. When that happened, it became advantageous for them to be able to support themselves on the seabed with their fins so they could raise their heads above the surface of the water.

G. Lundborg, *The Hand and the Brain*,
DOI 10.1007/978-1-4471-5334-4_1, © Springer-Verlag London 2014

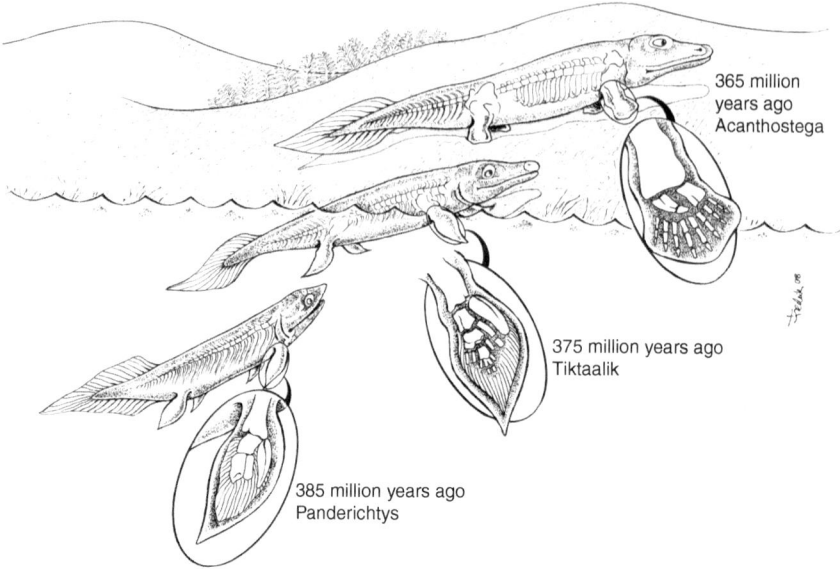

365 million
years ago
Acanthostega

375 million years ago
Tiktaalik

385 million years ago
Panderichtys

Fig. 1.1 Several hundred million years ago, when the earth's landmass consisted of marshes and swamps, many big fishes began a slow adaptation process towards life on land, involving the transformation of fins to extremities. One such example is *Tiktaalik*, representing a transitional phase between the fish *Panderichthys* and the tetrapod *Acanthostega*. The fins of *Tiktaalik* contained skeletal details reminiscent of a human arm/hand with an upper arm bone, two forearm bones and pre-stages of wrist and finger bones (Illustration: Fredrik Johansson)

In this transitional phase between life in the water and life on land, the fins of some fish species showed an obvious development towards an arm and a hand (Fig. 1.1). Perhaps can we regard these very early events as the first hints of a human hand and, accordingly, the first step towards a future human civilisation several hundred million years later.

A Fish Approaching Land

About 385 million years ago, at the end of the Devonian Period, several big fishes went through an adaptation process towards a land-based life, and several fish species with an amphibian-like appearance arose [2, 3]. One example is the *Panderichthys*, an alligator-like fish about 1 m long with eyes on top of a flattened head. Its fins had developed armlike skeletal structures with a shoulder blade (scapula), an upper arm bone corresponding to a humerus and two forearm bones corresponding to the ulna and radius.

However, it was still a fin, comprised of about 20 fine radial-like structures arranged along the periphery like a fan, giving it stability. The Panderichthys' head was firmly fixed to the body and could not be moved separately. The Panderichthys is considered to be the last fish in the development towards land animals with four extremities, the so-called tetrapods [4].

In 1931, a spectacular find was made on Ymer Island in Greenland. On the slopes of the Celsius mountain, a Danish expedition, led by the Swede Gunnar Säve-Söderbergh, found remnants of a 365 million-year-old fishlike skeleton showing four obvious extremity-like components. The specimen, named *Ichthyostega*, is still preserved in the Swedish Museum of Natural History in Stockholm together with another closely related fossil, *Acanthostega*, from the same area. The *Acanthostega* has four extremities and was clearly adapted for life on land. The two front extremities had well-developed arm bones with obvious finger structures; however, interestingly, there were eight digits on each of these rudimentary limbs. The extremities were stiff and nonmobile, more adapted for swimming or dragging itself along the seabed than for moving on land. The *Acanthostega* probably spent most of the time in the water even if it might have made short excursions up on land.

In the evolutionary process towards land-dwelling animals, it was quite a jump from the *Panderichthys* – the last fish – to the *Acanthostega*, a *tetrapod* characterised by extremities with fingers, adapted for a life on land. A component was lacking in the puzzle: a *missing link* between sea and land creatures.

Tiktaalik

At the end of the twentieth century, Ted Daeschler and Neil Shubin, both at the University of Chicago, and Farish Jenkins at Harvard decided to try to find such a missing link that could fill the gap between the fish *Panderichthys* and the tetrapod *Acanthostega*. They were well aware of the historical time window when such a transitional form might have developed – such fossil finds should exist in sedimentary layers from the late Devonian Period, about 375 million years ago. They decided to try to find these sedimental layers in the northern part of Canada, in one of the northernmost land areas on earth – the Arctic Ellesmere Island. A very adventurous and rigorous research project began, and its outcome was exciting. After an initial reconnaissance expedition in 2002, the team returned 2 years later and in a short time excavated several examples of a previously unknown species that perfectly fitted in the gap between a fish and a land-living four-legged *tetrapod* – a 'fishopod' – with all parts of the skeleton relatively intact [1, 5–8]. The find was named *Tiktaalik* after the local Inuit people's word for 'big fish' (Fig. 1.1).

Tiktaalik lacked the bony fusion that fixed the head to the body in the *Panderichthys* and instead showed a movable neck that allowed the head to turn in various directions. The well-preserved fossil skeletons had a short upper arm bone (*humerus*) and two short forearm bones corresponding to the *ulna* and *radius* in human arms. Distal to the forearm bones, there were even several rows of smaller bones that seemed to constitute an embryo of a wrist region as well as an early stage of fingers, where separate bone structures branched into several peripheral short rays [1].

The fossils were so well-preserved that the articular surfaces between separate bones were clearly visible. There was also a movable elbow joint with the ability to perform inward (pronation) and outward (supination) rotational movements of the 'forearm'. A striking find was a forestage of a movable wrist joint, which was lacking in previous fish species. *Tiktaalik* was hereby the first fishlike creature with a movable 'wrist joint' in its fins. Well-developed muscle origins, indicated by protrusions on the bone showing where muscles originated, proved that these fins were weight-bearing and could serve as support to raise the frontal part of the body above the surface of the water from the seabed, rather than supporting a walking action.

Thus, *Tiktaalik* was specialised for a rather extraordinary function: It was capable of doing push-ups. When we ourselves do push-ups, our hands are pressed against the ground, our elbows are bent and we use our chest muscles to move us up and down. In his book *Your Inner Fish* [1], Shubin describes how Tiktaalik's body was able to do all of this. The elbow was capable of bending like ours, and the wrist was able to bend to make the fish's 'palm' lie flat against the ground. Fins with a capacity to support the body were probably very helpful for a fish that needed to navigate the beds and shallows of streams and ponds and even to flop around on the mudflats along the banks [1].

Many of the structures and functions of our own limbs can be traced back hundreds of millions of years to the fins of fish living at that time. Our natural hand and arm movements use joints that first appear in the fins of *Tiktaalik* – they didn't exist before then [1]. It is a sobering thought.

The Swedish palaeontologist Per Ahlberg at the Evolutionary Biology Centre at Uppsala University believes that *Tiktaalik* is an animal with sufficient anatomical and physiological qualifications to leave marine life and adapt to life on land. However, he also emphasises that *Tiktaalik* was probably one of several evolutionary 'experiments' going on at the same time. Nonetheless, *Tiktaalik* has been regarded as a 'missing link' between fish and tetrapods [9, 10].

In summary, it seems that a preliminary sketch of a human hand already appeared in the fins of the big fishes about 370 million years ago, a period when many fish species underwent a development towards living on land.

Perhaps the possibility of slight pronation/supination as well as flexion/extension in the fins of *Tiktaalik* was a crucial factor in making the step from sea to land possible. These special movements ultimately became very important in the fully developed human arm and hand several hundred million years later: supination and pronation of the forearm as well as flexion and extension of the wrist joint are essential movements for many activities of daily life, for example, turning a key, handling a corkscrew, opening a can or turning a book over to read the back.

It is difficult to understand and define developmental stages that occurred several hundred million years ago, and new unexpected finds can certainly make it necessary to reappraise current concepts. A Polish-Swedish research team recently reported finding fossilised footprints of *tetrapods* in the Zachelmie stone quarry in Poland [11]. The footprints have been dated to about 375 million years, which is 18 million years older than the oldest known skeletal fossils of *tetrapods* so far. Finds like these may make it necessary to reappraise our current picture of the origin of the first land animals. These footprints indicate that big four-legged animals measuring three metres long existed as long ago as 375 million years.

It is a mystery why the *Acanthostega*'s eight fingers over time developed into five fingers in the human hand. At some time during the complex evolutionary process, the number of fingers was reduced to five. Perhaps one might also wonder how this has influenced the way we count. Our number system is based on tens – how could we count today if we had some other number of fingers?

References

1. Shubin N. Your inner fish: a journey into the 3.5 billion-year history of the human body. London: Pantheon Books; 2008.
2. Shubin N, Tabin C, Carroll S. Deep homology and the origins of evolutionary novelty. Nature. 2009;457(7231):818–23.
3. Hall BK. Fins into limbs: evolution, development, and transformation. Chicago: University of Chicago Press; 2007.
4. Zimmer C. At the water's edge: macroevolution and the transformation of life. New York: Free Press; 1998.
5. Daeschler EB, Shubin NH, Jenkins Jr FA. A Devonian tetrapod-like fish and the evolution of the tetrapod body plan. Nature. 2006;440(7085):757–63.
6. Shubin NH, Daeschler EB, Jenkins Jr FA. The pectoral fin of Tiktaalik roseae and the origin of the tetrapod limb. Nature. 2006;440(7085):764–71.
7. Pennisi E. Paleontology. Fossil shows an early fish (almost) out of water. Science. 2006;312(5770):33.
8. Holmes B. Meet your ancestor. New Scientist. 2006;(2568):35–9.
9. Boisvert CA, Mark-Kurik E, Ahlberg PE. The pectoral fin of Panderichthys and the origin of digits. Nature. 2008;456(7222):636–8.
10. Ahlberg PE, Clack JA. Palaeontology: a firm step from water to land. Nature. 2006;440(7085): 747–9.
11. Niedzwiedzki G, Szrek P, Narkiewicz K, Narkiewicz M, Ahlberg PE. Tetrapod trackways from the early Middle Devonian period of Poland. Nature. 2010;463(7277):43–8.

Chapter 2
The Hand, the Brain and Man's Travel in Time

Abstract The interplay between the hand and the brain in the evolution of man is a fascinating subject. Current knowledge, previously based only on fossil discoveries, has recently been substantially increased, thanks to advances in DNA technology. About 4.4 million years ago, hominins, about 120 m in height, had well-developed hands with opposable thumbs, but only rudimentary brains (*Ardipithecus ramidus*). Many parallel hominin lineages probably evolved and disappeared through the evolutionary process, with the survivors showing a dramatic increase in brain size over the subsequent millions of years. It is believed that bipedalism, together with well-developed hands and the emerging capacity for tool making, were important factors in this process along with such other factors as dietary shifts towards meat and marine food and an evolving capacity to use open fire to process and cook food. Our own species, *Homo sapiens*, probably emerged in Africa about 200,000 years ago and emigrated out of the continent about 50,000–60,000 years ago, although these time estimates might have to be shifted back in time due to ongoing re-evaluations of the mutation rate in the 'genetic watch'. Our early ancestors (*Homo erectus*) probably left Africa millions of years ago, migrating to Europe and Asia, evolving into Neanderthals in Europe.

An enormous time period – at least 350 million years – was required to allow the tetrapods, the first land creatures, to develop into early prehumans characterised by an upright posture and bipedal walking and to shape a human hand out of Tiktaalik's fin. There is not much fossil material available that can give us a clear picture of all developmental phases during this huge time period, and for a long time it has been unclear just when our earliest upright-walking ancestors appeared in the flatlands and forests of Africa. These early non-human primates, previously known as hominids, have more recently come to be designated hominins by most scientists.

The sensational 1974 discovery of Lucy, belonging to the species *Australopithecus afarensis*, surprised the world with a totally new insight – that 3.2 million years ago the northeast parts of Africa hosted early ancestors to us who walked upright on 2 ft but had very small brains with a volume of only 400 cc,

G. Lundborg, *The Hand and the Brain*,
DOI 10.1007/978-1-4471-5334-4_2, © Springer-Verlag London 2014

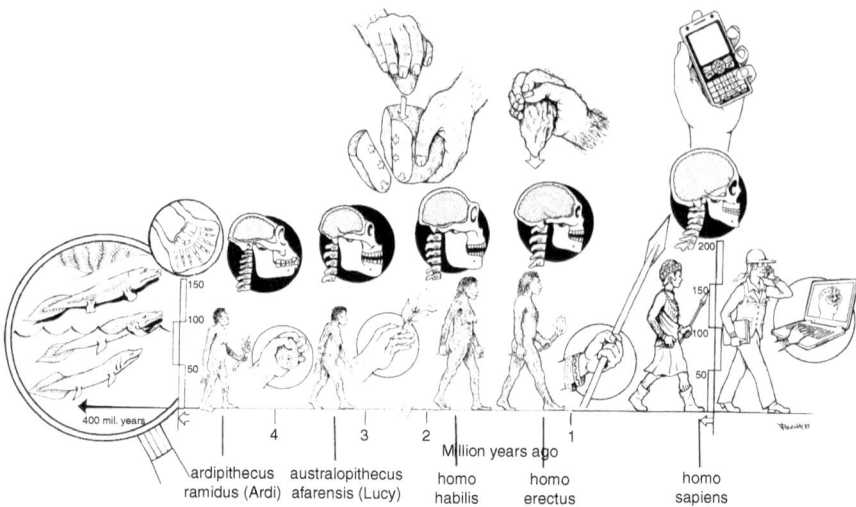

Fig. 2.1 Evolution of man from Ardi (*Ardipithecus ramidus*) to our own species *Homo sapiens*, illustrated on a timeline. Ardi was able to walk upright, but her big toes deviated somewhat, indicating an adaptation to also climbing in trees. Ardi's brain was very small, about one fourth of the brain volume of modern man. The hands of Ardi as well as the later Lucy (*Australopithecus afarensis*) were well-developed with long, slightly curved fingers adapted for gripping around tree branches, but also constructed to give a good precision grip. The thumb was opposable, meaning a good ability to grip around stems, to pick fruit and throw stones. The oldest known manufactured tools were long associated with *Homo habilis* who had found out how to strike off sharp flakes from the sides of stones for cutting and scraping purposes. However, recent discoveries have indicated that Lucy might also have been able to make simple stone tools. *Homo erectus* was able to make more advanced tools, called hand axes. In our own species, *Homo sapiens*, the hand and brain interact and cooperate in a way that allows the handling of advanced technical equipment like computers and smart phones. The drawing is very simplified since several parallel species might well have existed during the same time periods (Illustration: Fredrik Johansson)

approximately corresponding to a chimpanzee brain [1]. But these hominins were already equipped with very well-developed hands, largely resembling our own hands.

But an even greater discovery was presented to the world in October 2009 in *Science*. About 4.4 million years ago, that is, 1.2 million years before Lucy, the same region hosted an even earlier two-footed hominin: 'Ardi', of the species *Ardipithecus ramidus* [2]. Ardi was not only upright walking but showed, even at this early time, very well-developed hands [3]. Ardi's brain was even smaller than Lucy's – the volume has been estimated at only 350 cc.

With the discovery of Ardi in 2009, the evolution of man was suddenly re-evaluated, and the dating of the first known upright-walking hominin was shifted 1.2 million years back in time. Apparently, the first hominins in the human development line were upright walking more than 4 million years ago; they had very small brains but surprisingly well-developed hands – an interesting stage in the ongoing evolutionary process (Fig. 2.1). Scientists have speculated on how the brain could expand and increase its volume 3–4 times in a couple of million years, and many believe that the early well-developed hand and the early use of tools may have played an important and perhaps crucial role in this process. Together with several other factors such as a dietary shift towards more meat and cooked food [4, 5], the hand's abilities, possibilities and activities were important factors in the development of the brain and thereby also the development of consciousness, intelligence, creativity and the capacity to make tools. The free hands constituted a basis for a primitive sign language and with time also a spoken language. The ability to carry one's baby when individuals had to move added a new dimension to life. The hand made it possible to carry home fruits, roots and bags that one could share with other individuals in the group – cooperation among individuals was a prerequisite for survival, a sense of solidarity and with time also the formation of small groups and societies.

Human evolution and the origin of our own species, *Homo sapiens*, has inspired research primarily among palaeoanthropologists, biologists and cognitive scientists. Our knowledge, which was previously based exclusively on fossil finds, is today largely based on the results of molecular biological research, with a focus on DNA. Our current knowledge regarding the evolution of the hand and brain has been reviewed in several monographs and journals focusing on various aspects, for instance, Donald Johanson's *Lucy* [1], John Napier's *Hands* [6], Frank Wilson's *The Hand* [7], Peter Gärdenfors' *How Homo Became Sapiens* [8], Michael Corballis' *From Hand to Mouth* [9], Steven Mithen's *The Singing Neanderthals* [10] and Ian Tattersall's *Master of the Planet* [11] and *What makes us human*, *Scientific American*, *Special Collector's Edition* 2013 [12].

The Origin of Man: A Prestigious Story

The origin of man and the location of our earliest ancestors became a matter of prestige with considerable political impact, not least in terms of expressing a strong nationalism. The first fossil finds of early human-like beings were found in Asia, and consequently it was believed for a long time that Asia was the cradle of mankind. The first discovery was made in 1892 by Eugène Dubois who, during excavations on Java, found a prehistoric cranium with obvious human characteristics – 'Java Man'– who was long believed to be the missing link between monkey, ape and man. In 1929, a similar discovery was made by the Swedish scientists Otto Zdansky and Birgitta Bohlin, this time in the Zhoukoudian cave outside Beijing –'Peking Man'. Today we know that Java Man, as well as Peking Man, belonged to the species *Homo erectus*, 'the upright-walking man', an early member of the Homo family who had migrated from Africa to Asia a million years ago but who later disappeared in competition with *Homo sapiens* – 'the thinking man' – who migrated out of Africa much later.

However, in the western part of the world, people were quite confident that Europe was the cradle of mankind, and several finds of prehumans were presented from various parts of the European continent: skulls of Neanderthals were discovered in 1856 in the Feldhofer cave in Neanderthal in Germany.

The Neanderthals were long believed to belong to our own species, *Homo sapiens*. Today we know that the Neanderthals constituted a specific line within the Homo family and that they in fact were descendants from a much earlier species, *Homo heidelbergensis*, which existed about half a million years earlier and was also a common ancestor of our own species, *Homo sapiens*. The French reported about the Cro-Magnon man after finds of human-like skeletons in Cro-Magnon in 1868. Today we know that the Cro-Magnon man represented early immigrants from Africa to Europe about 30,000 years ago.

But from England there were still no reports of early human beings. A good illustration of the desperate search for early humans in England is the discovery of the 'Piltdown man', described by the British researcher Charles Dawson in 1912. He presented well-preserved fragments of a human-like skull: a lower jaw and a couple of molars that he had found in a gravel pit in Piltdown and that he felt proved that Britain was the cradle of mankind.

However, it was ultimately found that the Piltdown man was a clever bluff. The skull was composed of a contemporary human cranium and a jaw from a chimpanzee where the sharp teeth had been ground down to make them look more human. This discovery was regarded as the scandal of the century in the field. We never learned who the forger was, even if the man behind the discovery was strongly suspected.

Africa: Our Original Homeland

By tradition it was impossible for anyone to believe that Africa was the cradle of mankind, and it took a long time before such a concept could be accepted by the palaeoanthropologists. However, a series of fossil finds in the beginning of the twentieth century led to a total re-evaluation of the old concepts, and today it is generally accepted that Africa is the continent from which our early ancestors once emerged.

However, our evolution is not a long, unbroken line but rather a large number of 'evolutionary experiments' where several parallel developmental lines appeared and disappeared like an irregular bush with numerous sprawling branches, some of which end blindly and abruptly.

In 1924, a sensational finding in South Africa radically changed our view on the origin of man. In Taung, near the diamond city Kimberley, the Australian scientist and anatomist Raymond Dart found a skull with an appearance no one had seen before: an apelike cranium with obvious human characteristics like small canines but with a skull that could have hosted only a very small brain with a volume of 450 cc, about one-third of the brain of a modern human being. Strangely enough, the foramen magnum, the big hole in the skull, indicating a passage of the spinal cord into the skull cavity, was located at the bottom aspect of the skull, indicating an upright posture. The sutures of the skull bones were not yet fused, indicating that it was a child's skull. The find, which has been dated to 2.6 million years ago,

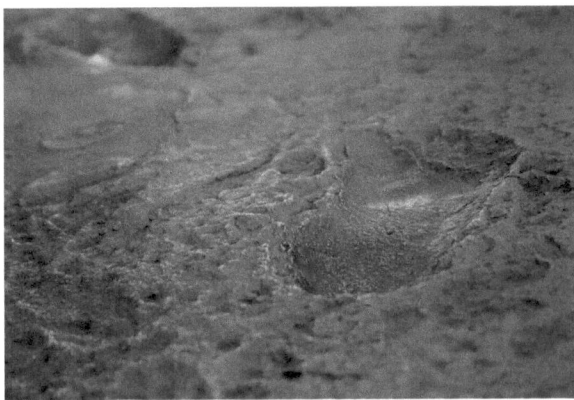

Fig. 2.2 A 3.7 million-year-old foot imprint of an *Australopithecus afarensis* from Laetoli in Tanzania. The picture shows a cast of a right foot with the big toe parallel to the other toes but slightly deviating – a primitive sign that is much more obvious in the earlier Ardi (*Ardipithecus ramidus*). The foot arch is obvious, and recent finds from Donald Johanson's research group show that Lucy walked like a modern woman (Photo: Gabriela Pichler)

indicated for the first time a possibility that man may have originated in Africa. The controversial discovery was much debated: who could believe that Africa was the cradle of mankind? Dart named his find *Australopithecus africanus* – the ape from the southern part of Africa. This and other similar finds from South Africa from the same time period indicated that our earliest ancestors were two footed and walked upright at a time when the brain was still very small – a totally new concept.

This observation was reinforced 50 years later when, in 1978, Mary Leakey made a sensational discovery in Laetoli, Tanzania – a 27-m-long track of footprints from upright-walking individuals who had walked over an area of flatland covered with volcanic ash that had solidified, preserving the footprints [13–15]. Probably two grown-up individuals and a child had moved across the landscape together. The footprints dated to 3.7 million years ago and seemed more or less identical to human footprints, with imprints of the arch of the foot and the big toes that (in contrast to monkeys living in trees) pointed straight forwards, parallel to the other toes (Fig. 2.2). The footprints, which indicated a human way of walking, have been associated with hominins of the type *Australopithecus afarensis*, among which 'Lucy' is the most well known.

Lucy

In 1974, Donald Johanson and his associates made a sensational find in a dried river bed in the Hadar desert in Ethiopia – a large number of skeletal fragments from one single prehistoric individual. The group found pieces of bones from the upper and lower extremities, vertebral column, ribs and pelvis together with sparse hand bone fragments. For the first time, the finds made it possible to put together a fairly complete skeleton of one unique individual: a prehuman who lived 3.2 million years ago. The find was named *Australopithecus afarensis*, 'the southern ape from Afar' [1].

In the research camp, the staff worked for several days putting together the parts of the skeleton, a tedious process accompanied by the Beatles tune 'Lucy in the Sky with Diamonds' – a good reason to name the find *Lucy*. Together with several similar finds from the same area, it now became possible to approximately reconstruct a *hominin* – a bipedal prehuman – who had lived in the area several million years ago. It has been concluded that Lucy was upright walking, but later analysis of fossilised scapulae supports the hypothesis that her locomotor repertoire included a substantial amount of climbing [16]. Lucy measured little more than 1 m in height and had a very small brain size of about 400 cc, like a chimpanzee's brain. Today Lucy is preserved in the National Museum of Ethiopia under the domestic name Dinknesh (wonderful) [10].

How well-developed were Lucy's hands and what could she perform with them? The hands of early tree-living apes were adapted for a life in the trees: the fingers were long, slender and a little bit curved to create an effective grip around tree branches, while the thumb was rudimentary, short and fairly nonmobile. Many of these finger characteristics were found in Lucy's hand, but in contrast to earlier tree-living apes, the hand had a well-developed and very mobile thumb with 'opposition' abilities; the thumb could be moved in a counteraction across the palm of the hand against the other fingers [7, 10, 17–20]. A prerequisite for this function is a saddle-shaped joint at the base of the thumb. Thus, Lucy probably had very good gripping function in her hand with a high degree of precision in fine motor movements. However, Lucy probably lacked the capacity to cup the hand, which is a crucial ability of the human hand for really good hand function. It was not possible to make a full and detailed analysis of Lucy's hand as there were only sparse bone fragments from a few fossil hands to study.

Among several important characteristics in Lucy's arm were the ability for flexion and extension in the wrist joint and the capacity to rotate the forearm inwards and outwards (pronation/supination) [21]. These forearm abilities were probably a prerequisite for throwing stones with precision and for gripping and using primitive weapons for defence and in fights. The mobility in the forearm, wrist and thumb are qualities that would make it possible for the hand to make and use tools, provided that the mental resources were sufficient. But the perception has been that Lucy, with her small brain, did not have this capacity – the hand was certainly developed into an instrument of great precision, but there was still no effective superior control organ – a well-developed brain. No handmade tools have been found from Lucy's time period, but in August 2010, *Nature* presented new finds indicating that hominins from Lucy's time actually seem to have used simpler stone tools to scrape and discard meat from carrion bones and to crush bone to access the bone marrow [22]. It is difficult to determine whether these stone tools were spontaneously found stones of suitable shape in the immediate vicinity or if they have been purposely made by hand. The tools were discovered only 200 m from where a fairly complete skeleton of a girl belonging to the species *Australopithecus afarensis* – Lucy's baby – was found [23] in Dikika, just 4 km from where Lucy turned up. These new finds have attracted much public attention since they suddenly shift the established time table for use of tools by our early ancestors back about 800,000 years. According to current belief, so far the earliest use of stone tools to handle the meat and bone of animal carcasses took place 2.5 million years ago, a dating that is based on finds in Gona, Ethiopia.

'Lucy's baby' was named Selam – meaning 'peace' in several Ethiopian languages – in hopes of encouraging harmony among the warring tribes of Afar [24]. The skeleton, judged to be that of a 3-year-old girl, consisted of a virtually complete skull, the entire torso and parts of the arms and legs. Selam had legs built for walking and long curved fingers built for climbing, indicating that she was suited to an arboreal existence but also well adapted for walking on the ground.

Recent fossil finds in Ethiopia have shown that Lucy might have been just one of two or more parallel developing lines existing at the same time. Based on bone specimens from a foot, it has been concluded that another species, able to walk upright but also spend much time in trees, existed during the same time period as Lucy [25].

Ardi: Our Earliest Ancestor

For a long time Lucy was regarded as our 'first mother' in the development towards our own species *Homo sapiens*. But with the 2009 presentation of Ardi – *Ardipithecus ramidus* – it became necessary to re-evaluate and reappraise the earliest development of man. Tim White, professor of Integrative Biology at the University of California in Berkeley, and his team spent 15 years analysing and reconstructing skulls, teeth, pelvic bones, hands, feet, arms and lower extremities from at least 35 individuals who lived in northeast Ethiopia 4.4 million years ago, not far from the area where Lucy was found in 1974 [2, 3, 26]. Among the finds were several very well-preserved skeletons, and for the first time it became possible to reconstruct in detail complete skeletons of one of our earliest ancestors.

The summarised finds were presented in 11 articles in the same issue of *Science*. Ardi's skull hosted a brain with a volume of 300–350 cc [27], that is, a brain considerably smaller than Lucy's. The anatomy of Ardi's pelvis, lower legs and feet indicated that she had an upright posture and an upright way of walking [28]. She measured 120 cm in height and her weight was about 50 kg. The feet were big and well adapted for upright walking, but Ardi also had a deviated big toe, which was useful for life in the trees [29]. Thus, she was probably adapted for a life in trees as well as on the ground. Her canines were small and not at all as big as those of chimpanzees, indicating a more human-like and less aggressive character.

Finds of wood fragments and seeds indicate that Ardi lived in a landscape dominated by sparsely growing fig, elm and palm trees. A large number of fossils

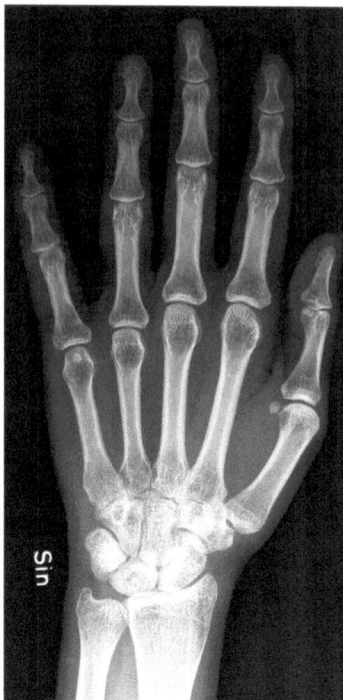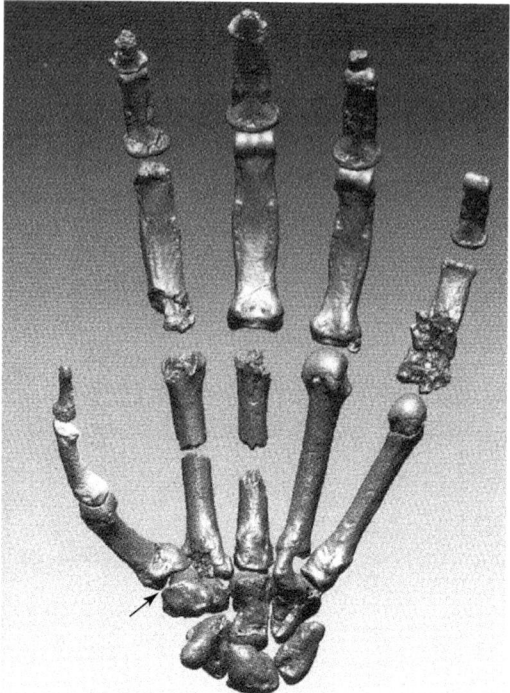

Fig. 2.3 Ardi's (*Ardipithecus ramidus*) hand skeleton (*right*), dated 4.4 million years ago, compared with the hand of a modern human (*left*). There are many similarities between Ardi's hand and that of a modern human. Ardi's thumb is certainly much shorter, but at the base of the thumb is a 'saddle joint' (see the *arrow*), like in modern man, that makes it possible for the thumb to touch the other fingers and to move in several planes. This is a prerequisite for good hand function with high precision (From Lovejoy et al. [3]. Courtesy of the American Association for the Advancement of Science)

from animals such as porcupines, hyenas, bears, rhinoceroses, pigs and elephants were found in the same area. Ardi was probably an omnivore with a preference for fruits, mushrooms and small animals [29].

Ardi's arms were long and slender and reached knee level [3]. Her hands were very mobile with an anatomy adapted to a life in the crowns of trees: a very flexible wrist and long, slightly curved fingers with a good capacity for effective gripping around branches. But Ardi's hands also had good precision capacity and good fine motor functions, exhibiting several of the characteristics also found in Lucy's hand 1.2 million years later (Fig. 2.3). Ardi's thumbs were well-developed and opposable, indicating a very good hand function [2, 3]. The anatomy of the hand bones showed that these individuals did not support the weight of their bodies on their knuckles when they walked: they were bipedal and moved in an upright position.

With the discovery of Ardi, our views on our earliest ancestors changed dramatically. It was long believed that the human line had passed a 'chimpanzee phase', but Ardi does not show any similarities to chimpanzees; instead she was characterised by several primitive features reminiscent of even earlier non-human primates. This means that Ardi must have belonged to a species that appeared soon after the stage about 6–7 million years ago when the developmental line split into two branches: one that further developed towards chimpanzees, while the other constituted a 'human line' leading to *Ardipithecus ramidus* and ultimately to our own species, *Homo sapiens*.

Ardi was perhaps our *original ancestor* – upright walking with a well-developed hand and a long and moveable thumb, but still with a rudimentary brain. However, Ardi may have belonged to only one of several parallel developing lineages, and today no one can say for sure which of these lines ultimately led to *Homo sapiens*. Either way, the foundation was laid for a continuous interaction between hand and brain over the ensuing several million years, with the brain developing towards capacities like abstract thinking, tool making and, with time, communication via symbols, pictures and ultimately a sign or spoken language.

An article in *Science*, published on 18 December 2009, proclaimed the discovery of Ardi 'The Breakthrough of the Year' [30].

Homo habilis: The 'Handy Man'

Mary and Louis Leakey carried out extensive excavations for several decades in their search for human fossils on the shores of Lake Victoria and in Olduvai Gorge in Tanzania. They found a large number of fossils from several types of extinct animals. But most of all they were hoping to find traces of our earliest ancestors. On 17 July 1959, the search was successful. At the time Louis was ill, so Mary went on a tour around the area herself. Suddenly she sighted some bone fragments that looked interesting, and upon closer inspection they turned out to be two big canines of an early *hominin*. This was the first find of *Australopithecus* (*Zinjanthropus*) *boisei*, which has been dated to about 1.75 million years ago [31]. Anthropologists named the find *Zinj*. The very powerful molars indicated a diet consisting of fruits and nuts – consequently the hominin got the nickname *nut-cracker man*. The find, which received much attention, resulted in renewed financial support from the National Geographic Society, making further excavations in Olduvai Gorge possible. Still, the Leakeys were hoping for fossils from our own genus *Homo*.

In 1962, the Leakey couple reported new spectacular finds in Olduvai Gorge. This time they reported four skulls of less primitive appearance, with a larger skull cavity that could host a larger and more developed brain than earlier finds of *Australopithecus* – about 640 cc. They also found a lower jaw and several hand and foot bones. At the same location, they found a simple type of stone tool: fist-sized round stones that had been sharpened by striking several sharp flakes from the sides of the stone. The sharp edge of these 'pebble choppers', as well as the edges of the flakes, was probably very suitable for processing meat and cutting the skin of animals, while the larger stone would easily crush bones to access the bone marrow. The Leakeys named this manufacturing of simple stone tools *Oldowan*.

The Leakeys felt that they had now found a hominin with a more well-developed brain than earlier species. They believed they had found the first individuals of a new genus – *Homo* – and they named the find *Homo habilis*, 'the handy man'.

The finds were presented 1964 in *Nature* with the prestigious title 'A new species of the genus *Homo* from Olduvai Gorge' [32]. Louis Leakey strongly argued that he had now found a fossil of the first real human being. It was felt that *Homo habilis*, dated 2.4–1.6 million years ago, probably had its origin in some species of *Australopithecus*. *Homo habilis'* weight has been estimated at about 24 kilos.

The hand of *Homo habilis* was characterised by long, slightly curved finger bones where the most distal bones (end phalanges) in the fingers were broad and human-like. Like Ardi and Lucy, *Homo habilis* had a relatively well-developed thumb that was opposable, an important property giving the hand great precision and a good gripping function [6, 33–35]. Even if Ardi, 2 million years earlier, already had a well-developed hand, the researchers argued that *Homo habilis* was the first hominin who – with its more developed brain – possessed sufficient capacity for planning and abstract thinking to fully utilise the hand in making tools. However, these conceptions are now being questioned, since evidence has recently been presented showing that the earlier species, *Australopithecus afarensis*, might also have been able to make simple stone tools (see also Chap 3).

Homo erectus: The Upright-Walking Man

Lucy and her ancestors were probably very ape-like, and *Homo habilis* probably did not look much different. The big change came with the succeeding *Homo erectus*, who had a much more human-like appearance. *Homo erectus* was believed to exist 1.8 million–300,000 years ago. While earlier species had relatively long slender arms, short bones and a curved posture, *Homo erectus* was more human-like and tall with longer legs and more upright walking. Many of modern man's typical characteristics were present in *Homo erectus* [10]. The brain had undergone an explosive development, and the posture of the body was more like modern man. 'The boy from Nariokotome' in Kenya is one of the most well-preserved fossils of *Homo erectus*, probably 1.6 million years old. The boy from Nariokotome never reached puberty, but as a grown-up he would probably have been more than 180 cm tall and weighed about 68 kilos. This means that our ancestors, in a period of just 100,000 years, underwent an explosive development and became almost three times as big as the small *Homo habilis*.

Recently discovered footprints of *Homo erectus* in sedimentary layers close to the city of Ilere in northwest Kenya, probably about 1.5 million years old, show that the feet of these individuals were anatomically more or less identical with modern man's feet, with high arches and big toes pointed straight forwards, parallel to the other toes [36, 37]. The tracks of the footprints also show that *Homo erectus* had an efficient way of walking, corresponding to modern man, with the heel obviously lifting while the front of the foot was still on the ground. The footprints indicate a more developed anatomy in a 'modern' direction as compared to the footprints of *Australopithecus afarensis*, which had previously been found in Laetoli, Tanzania, and are estimated to be 3.7 million years old. Probably the well-developed feet of

Homo erectus contributed to an improved walking capacity, making the very long migrations characterising this species possible: Many of these individuals emigrated from the African continent.

Fossils of skulls show that *Homo erectus* had small canines and a considerably larger brain volume than its predecessors, about 1,000 cc. Researchers found more advanced tools close to *Homo erectus* fossils than those found near older fossils of hominins: so-called hand axes, that is, oval or pear-shaped stones where flakes had been symmetrically knocked off from both sides to achieve a sharp cutting edge. These activities were named *Acheulian* (after Saint-Acheul, a suburb of Amiens in northern France where some of the first finds were made). Acheulian hand axes were processed with stones, horn, bone and wood to get their final form and sharpness. They were a general-purpose tool that could scrape, cut and stab skins of wild animals preparatory to skinning [6]. Tools of this type have been found in Africa, Asia and Europe. It is generally agreed that the hand axe was an extension of the pebble chopper associated with the Oldowan culture. From originally being very primitive in appearance, over time hand axes became increasingly sophisticated, probably reflecting a corresponding development towards more advanced thinking and an improved aesthetic sense of form and decoration.

The Acheulian culture probably began the art of hunting, and researchers believe that groups of early types of *Homo erectus* migrated from Africa looking for quarry and new hunting grounds very early, perhaps beginning 1.8 million years ago. Over various time periods, there were most likely waves of emigrants moving out of Africa. Recent finds from southern Spain show that hand axes were used in Europe about 900,000 years ago [38]. Tools of Acheulian type, estimated to be 780,000 years old, have also been found in the area around the Red Sea. In Asia, *Homo erectus* gave rise to 'Java Man' and 'Peking Man'. Finds indicate that *Homo erectus* lived in the Zhoukoudian area outside of Beijing about 770,000 years ago at a time when this part of the world was experiencing an arctic climate [39].

Homo erectus was characterised by a capacity for cooperation, social interaction, innovative thinking, good memory function and the ability to use and teach others to use fire. The oldest known use of fire has been dated to 1.8 million years ago; small balls of clay were found that had been burnt at a temperature of about 400 °C, which is a temperature corresponding to a camp fire.

The use of open fire was probably linked to abstract thinking, social interaction and a capacity for communication, possibly even a primitive verbal language. Fire gave heat and light and was also important for cooking meat, a process of importance to optimise the energy uptake from the food and to facilitate its digestion in the alimentary tract, a factor which might have been of major importance for the increase in brain volume. Fire had the important social function of bringing the group together and keeping insects and predators away. Much later in development, fire was probably also used to heat stones so that flakes could more easily be knocked off. At Pinnacle Point on the east coast of South Africa, 164,000-year-old remnants from such a heat treatment of siliceous pebbles have been found [40].

Homo erectus may have existed in several forms. One of those, *Homo ergaster*, is regarded by many scientists as the original African form, making *Homo erectus* a later form that emigrated to Asia [10]. Recent fossil discoveries indicate that an additional species, *Homo rudolfensis*, might have existed parallel to *Homo erectus* and the earlier *Homo habilis*. This idea is based on a recent find of a fossil lower jaw close to Lake Turkana in Kenya, dated from 1.78 to 1.95 million years ago. Thus, according to the discoverer, the palaeoanthropologist Meave Leakey, three different *Homo* species probably existed during the same time period, about 2 million years ago [41]. Leakey believes that there were at least two parallel evolutionary lineages at the early stage of *Homo* to which the species *Homo habilis* and *Homo rudolfensis* are assigned. However, whether either of these two lineages was an ancestor of *Homo erectus*, let alone of modern humans, remains uncertain.

The Neanderthals

The Neanderthals, *Homo neanderthalensis*, lived in Europe and Asia about 350,000–30,000 years ago [42–44]. They dominated the area from Portugal to Uzbekistan for several centuries. The Neanderthals are our closest relatives, even though it is documented that *Homo sapiens* is a specific species within the genus *Homo* and not a direct descendant of the Neanderthals. The first fossil of a Neanderthal individual was found in 1830 in Belgium, but initially the importance of this find was not understood. Not until a similar find was made in 1856 by mine workers in Neanderthal in Germany did scientists understand the importance of the discovery, though it was first argued that the bones were the remains of a malformed Russian Cossack who had died during the liberation war against Napoleon in 1813–1814.

Our knowledge of the constitution of the Neanderthals and their life conditions was originally based on skeletal remains, but in recent years DNA research has provided much new knowledge [42, 45, 46]. The Swedish DNA scientist Svante Pääbo, active at the Max Planck Institute for Evolutionary Anthropology in Leipzig, has long been working to map the complete Neanderthal genome. The results show that the human line and the Neanderthal line separated about 500,000–800,000 years ago when both lines split off from a joint ancestor, *Homo heidelbergensis*. Until recently it was difficult to find any genetic traces of Neanderthals in modern man, and the opinion has long been that the two species *Homo neanderthalensis* and *Homo sapiens* did not intermingle with each other, or at least that there was no interbreeding, during the 15,000 years that both species coexisted. Recently this opinion has been revisited by Svante Pääbo and his research group at the Max Planck Institute. In an article published on 7 May 2010 in *Science*, Svante Pääbo and his team showed that 2–4 % of the genes of modern man are shared with Neanderthals [47, 48]. The conclusion is based on an analysis performed on three 38,000-year-old Neanderthal bones from the Vindija cave in Croatia. The Neanderthal genetic contribution is found in people from Europe, Asia and Oceania.

Thus, it seems that the two human species did intermingle to a small extent and that they even produced children. This means that people in Europe, Asia and

Oceania have a fragment of Neanderthals in their genes; 'they are not totally extinct. In some of us they live on a little bit', says Professor Pääbo. So, when we look at ourselves in the mirror, we catch a glimpse of a long-extinct relative – the Neanderthal. It is certainly a staggering and thrilling insight.

The Neanderthals had somewhat larger brains than modern man. They had a robust body with prominent brow, long arms, short legs, a broad thorax and a broad pelvis. Their gait was different from ours, stiffer with greater rotation of the hips when walking [11]. Neanderthal faces bore large noses, and their cheekbones retreated rapidly at the sides. Analysis of Neanderthal DNA samples suggests that they possessed an inactive version of a gene affecting skin and hair colour – apparently these individuals would have possessed pale skin and red hair [11]. The gene FOXP2, linked to language capacity, was found in Neanderthals as well as modern man, indicating that the Neanderthals may have possessed a certain capacity for verbal communication.

The hands of Neanderthals were large and powerful with broad fingertips [49, 50]. The hand muscles were voluminous and well-developed. Some researchers proposed that the hands of Neanderthals did not have the same good hand precision ability as modern man [49], but careful analysis using modern 3D X-ray techniques has shown that the anatomy of the thumb and fingers was more or less identical to modern man. The opposability of the thumb was the same, and the Neanderthals probably had an effective precision grip between the thumb and index finger, just like ours [51]. Thus, there is no reason to believe that the hand function of Neanderthals was inferior to the hand function of modern man.

The Neanderthals were true cavemen and hunters, and they developed an effective technique for cutting up their quarry and using the fur for clothing, a prerequisite for survival in the cold climate. They were good hunters using advanced types of spears – in fact their progenitor, *Homo heidelbergensis*, was able to add shafts to sharpened stone points, creating spears [52]. They developed effective cutting tools as well as various types of scrapers to clean up the animals' skins. They used hand axes and various stone tools with cutting edges.

It has been proposed that the Neanderthals had inferior intellectual and physical qualities when it came to competing with *Homo sapiens* who later invaded the areas. The result was that the Neanderthals became extinct and disappeared about 25,000–30,000 years ago. Earlier it was not possible to demonstrate any signs of abstract thinking expressed in artistically shaped tools, personal ornaments or decorations. But recent finds indicate that the Neanderthals did, in fact, manufacture ornaments to adorn themselves. The archaeologist João Zilhão from the University of Bristol has recently discovered 50,000-year-old coloured cockleshells in two caves located in southeast Spain [53, 54]. Cockleshells with prepared holes were coloured with yellow and red pigment fetched from about 5 km from the place where they were found. According to Zilhão, the finds indicate that the Neanderthals had the same capacity for symbolic thinking and creativity as modern man.

Several researchers maintain that there are no conclusive signs that the Neanderthals were less intelligent than modern man [55]. They lived in caves where special areas were reserved for various activities like handling meat and making stone tools. There were special sleeping areas and specific areas reserved for refuse.

Fire played an important role in the life of the Neanderthals. In a series of caves about 100 km outside Barcelona, researchers have found obvious marks indicating fireplaces that apparently were kept in good shape and regularly cleaned of burnt remnants [56].

Neanderthals and individuals belonging to our own species coexisted for about 15,000 years. During the last period before the disappearance of the Neanderthals, they retreated to remote mountain areas in southern Spain, Croatia, the Crimea and the Caucasus. They probably had a hard time surviving in the increasingly colder climate that began in Europe about 50,000 years ago and that ultimately led to the last glaciation about 25,000 years ago. The last Neanderthals probably lived in caves in the Gibraltar cliff about 28,000–24,000 years ago [55].

The Hobbit from Flores: *Homo floresiensis*

In October 2004 a sensational palaeontological find was reported: a brand new species of the genus *Homo* [57]. On the Indonesian island of Flores east of Bali, a surprising discovery was made: the remains of eight individuals, less than 1 m in height, with small skulls whose brain cavities could host small brains of about the same size as Lucy's (about 400 cc), who had existed up to 13,000 years ago. It was proposed that the finds, dated to 18,000 years ago, belonged to a new species of the genus – *Homo floresiensis*.

However, the finds immediately became very controversial. Some researchers were of the opinion that *Homo floresiensis* in fact was a variant of our own species *Homo sapiens* with a small and undeveloped brain – a microcephaly. However, very strong evidence has been presented indicating that in fact they are more closely related to a separate species of the genus *Homo*, probably a descendent of *Homo erectus* who migrated eastwards from Africa millions of years ago and may have arrived in Flores about 900,000–800,000 years ago [58–62]. There are also theories that *Homo floresiensis* in fact descends from the *Homo erectus* predecessor *Homo habilis*, something which, if true, may indicate an even earlier migration out of Africa. *Homo floresiensis* has been called 'the hobbit' after the hobbit characters in Tolkien's *Lord of the Rings*. Despite their small brains, these individuals manufactured advanced stone tools like knife blades as well as barbed spearheads.

Matthew Tocheri, an expert on development of the human wrist joint, made an interesting observation regarding the anatomy of the bones in the wrist joint of *Homo floresiensis*. The trapezoid bone, *os trapezoideum*, a small bone in the wrist joint that, in modern man, is situated between the scaphoid bone and the metacarpal bone of the index finger, is small, rudimentary and obliquely positioned in *Homo floresiensis* [63]. The anatomy of this bone corresponds exactly to the corresponding bone in chimpanzees and the hominins in the *Australopithecus* species, but does not show any similarities to modern man. Even the appearance of the foot skeleton indicates that we are dealing with a primitive *Homo* species different from *Homo sapiens*. According to Tocheri, the anatomy of the hand bones as well as the foot bones in *Homo floresiensis* strongly indicates that we are dealing with a separate and solitary species in the genus *Homo* [64].

A Little Finger from an Unknown Prehistoric Relative

The details of the early development of man are still unknown. New finds make it necessary to constantly re-evaluate current perceptions. New DNA technology makes it possible to map the DNA of fossil material in a way that was not possible before. An example of this is how a fragment of a fossil bone from a little finger can reveal the existence of a completely unknown species of the genus *Homo*. In the summer of 2008, Russian scientists found fragments of a little finger bone in the Denisova cave in southern Siberia [65, 66]. It was supposed that the finger came from some of the Neanderthals who had lived in the cave 30,000–48,000 years ago. But when Svante Pääbo and his associates performed a DNA analysis of the bone tissue, there was a big surprise. There was no conformity between the DNA from the little finger and that of Neanderthals or *Homo sapiens* who also lived in the area at the same time [65, 67]. The conclusion was that the finger, which came from a woman (the 'X woman'), must have come from a previously unknown extinct species of the genus *Homo* who might have emigrated from Africa millions of years ago. It is a fascinating thought that 40,000 years ago Central Asia was populated not only by humans belonging to our own species and Neanderthals but also by individuals belonging to a third, hitherto unknown hominin lineage, which has been named Denisovans after the Denisova cave. No one can tell today whether individuals of these different species communicated with each other and how peaceful possible meetings might have been. However, there are data indicating that the Denisovans interbred with ancestors of modern-day Melanesians. Svante Pääbo and his associates recently showed that the Denisovans share 4–6 % of their genome with today's inhabitants of *Papua New Guinea*.

Another previously unknown early relative to us was recently presented in *Science* [68–70]. Matthew Berger, the 9-year-old son of the palaeoanthropologist Lee Berger at Witwatersrand's University in Johannesburg, found some skeleton fragments behind a stone. At first they were believed to be remnants from an antelope.

But after a closer look at this and similar finds from the Malapa cave north of Johannesburg, it was understood that the finds instead derived from a hominin, showing some primitive characteristics typical of *Australopithecus afarensis* (the same species as 'Lucy') but also showing characteristics typical of more well-developed individuals in the genus *Homo*. The find, which was dated to 2 million years ago, was named *Australopithecus sediba* (sediba is the Sesotho word for 'fountain' or 'wellspring') – the original root of mankind. *A. sediba* was about 1.3 m in height. In comparison with Lucy, the teeth were smaller and the bones longer.

The size of the brain was at least 420 cc, about the same as Lucy's. The anatomy of the pelvis resembled the much later *Homo erectus*.

It was first supposed by the discoverers that *A. sediba* represented a previously unknown *Australopithecus* species. But *A. sediba* shows more similarities to early *Homo* species than any other known *Australopithecus* individual, and it has been proposed by the discoverers that *A. sediba* could be the ancestor of *Homo*.

Most probably there are more surprises around the corner when it comes to the origin and development of man. DNA technology has opened up brand-new possibilities for identifying fossil material where a reliable classification was previously impossible. It is probably unlikely that very old DNA material is still preserved in warm areas of the earth, but several highly interesting archaeological areas still remain to be excavated in colder climate zones [65].

How Did We Become Bipeds?

For 50 million years, our ancestors lived as apes in the trees of tropical forests: the origin of man is among the big African apes.

About 15 million years ago, at the dawn of human development, some type of tree-living ape on the African continent descended from the trees to live on the ground. Probably the *orangutan* split off from the common developmental line 10–30 million years ago, while the gorilla split off the line 7–10 million years ago. The remaining common line split into a chimpanzee line and a *Homo* line at some time 5–6 million years ago. But still the genetic difference between chimpanzee and man is surprisingly small: our genome is at least 98 % identical [71].

A major event in the development of mankind was when some of our prehistoric ancestors once left the four-foot stage and became an upright-walking *biped* [72, 73]. The arms and the hands were no longer needed to support the body, but could be used for completely other purposes. This change in hand and arm function dramatically influenced the species' continuing development. But when did our prehistoric ancestors stand up to become *bipeds*? In 2000, there was much public attention around 'Millennium Man' – *Orrorin tugenensis* – who was discovered in northern Kenya by a French-Kenyan research team led by Brigitte Senut from the Musée National d'Histoire Naturelle in Paris. Millennium Man was dated to 6 million years ago, and the discoverers felt that they had now found the forefather of us all – the very earliest upright bipedal man [74]. However, the conclusion was based only on scattered skeletal pieces, including a femur whose appearance indicated an upright gait. Several researchers hesitated to interpret these finds and have put forward the opinion that *Orrorin*, who probably spent most of his time in the trees, indeed was an ape rather than an upright-walking hominin.

Among early hominins with firm indications of *bipedalism* and upright gait, *Ardipithecus ramidus* – 'Ardi' – plays an especially important role. As judged from the appearance of the pelvic bone, the lower extremities and the feet, Ardi was definitely bipedal. Another early bipedal hominin is *Australopithecus anamensis*, who, like Ardi, was found in the Awash region in Ethiopia and in northern Kenya. *Australopithecus anamensis*, who was dated to 4.2–3.9 million years ago, has been regarded as an intermediate species between *Ardipithecus ramidus* and the later bipedal *Australopithecus afarensis*. Thus, we can trace the upright gait back more than 4 million years in the history of evolution. Bipedalism occurred long before the brain developed to a larger volume. It seems that our ancestors at some period far back in time were under a strong selective pressure with a priority on bipedalism rather than intelligence [8]. But the driving force behind the upright body position is still a mystery. There are several different hypotheses.

The Savannah Hypothesis

According to the 'savannah hypothesis', the upright gait is associated with profound changes in climate and vegetation, including the transformation of forests into more open land, forcing the hominins out on the savannahs, into open land. By standing and moving in an upright position, these hominins could scan the savannah to spot carrion and quarry as well as dangerous predators. They had to move across long distances in open terrain to find food, and several researchers believe an upright gait made it possible to do so without fatigue. Thus, bipedalism has been regarded as a more effective way of moving than using four legs. Having their arms free made it possible to carry babies as well as food and various items. Cooperation among individuals in the group became necessary for protection against predators and to make hunting more efficient. Standing on two legs made it possible to pick fruit from trees and bushes with both hands. In addition, an upright-walking posture minimised the negative effects of the strong sun since the whole body was no longer directly exposed to the sun.

The early prehumans were capable of hunting and running after quarry over long distances until the quarry became exhausted, had to rest and thereby was easy to kill. The free arms allowed signalling and a kind of sign language, which improved the ability to communicate, even over great distances, before there were sufficient physiological and intellectual conditions for an articulated verbal language.

However, today the savannah hypothesis is regarded by most scientists as antiquated, and the arguments supporting the theory are irrelevant in many ways. It is doubtful whether bipedalism would be more effective than using all four extremities when it came to quick movements; all animals capable of running at high speeds use four extremities. It is certainly true that free hands are important to reach fruit and for carrying babies; however, it is well known that this is also characteristic of four-footed chimpanzees.

Perhaps the strongest argument against the savannah hypothesis is that man in fact did not acquire the upright position on savannahs but rather in a forest environment. Studies of animals and vegetation around areas associated with early hominins indicate that these individuals primarily lived in a forest environment. Strong front teeth and thick enamel shows that the primary food was fruit with elements of seed and roots, indicating an environment of rich vegetation. In addition, more or less all finds of early hominins have been made close to lakes and rivers, often in sedimentary layers close to water. These finds indicate that our ancestors lived close to water in an environment with rich vegetation. This does not preclude their spending spent time in a nearby savannah environment, but the savannah was probably not their primary living environment.

The Aquatic Ape Hypothesis

The aquatic environment may have played a big role in the development of our earliest ancestors, both with respect to development of the brain and the upright gait. About 5 million years ago, vast areas of northeast Africa were flooded and under water. Large bays of the sea invaded the land and a large number of islands were created. Isolation in an archipelago with forests drying out may have forced the early hominins to adapt to a water environment. Several researchers have also felt that our ancestors lived very close to or perhaps even in the water – like 'aquatic apes' – and that this may have stimulated an upright gait. The aquatic ape hypothesis was popularised in 1995 by the archaeologist Philip Tobias, who proposed that *Australopithecus* individuals may have used the trees for escape and for sleep, but that they otherwise lived very close to a water environment. The theory was originally proposed by the marine biologist Alistair Hardy in 1960 and by Elaine Morgan in several books [75, 76].

We still have many traits related to living in an aquatic environment, for instance, our lack of fur, our subcutaneous fat, our ability to voluntarily regulate our breathing and the fact that our noses are shaped to allow us to go underwater without it invading our respiratory tracts. In our very thin coating of hair, the individual hairs are oriented in a direction that allows water to stream along a swimming body. In addition, we have a high occurrence of haemoglobin in our red blood cells resulting in a high capacity for oxygen saturation in the blood, in many ways identical to several diving animals. Even the fully developed *Homo sapiens* are used to water from the embryonic stage: during embryonic life we exist in water for 9 months, and after birth babies are capable of staying under water for short periods. Even in the adult stage, we enjoy a water environment – at least during holidays!

The water environment would have allowed a diet consisting of mussels, shells and crustaceans – a diet rich in proteins that may have contributed to development of the brain. The buoyancy of the water would have made it easier to wade out to the mussel banks in an upright position, thus possibly stimulating the upright gait.

It has been proposed by many researchers that life in the marine environment contributed to human characteristics in several ways. For instance, it has been argued that our ancestors in the marine environment might have difficulties communicating with body language but that breath control, acquired during marine life (rapid inhalation and slow exhalation), may have created conditions for communication using sounds – a first hint of a verbal spoken language.

Did Bipedalism Develop While We Were Still in the Trees?

There are theories that the upright gait may have occurred very early in history, when our ancestors still were living in trees. If so, our ancestors simply transferred the upright gait to the ground.

These theories have appeared from observations of orangutans on the Indonesian island of Sumatra. Biologist Susannah Thorpe from Birmingham University observed orangutans living in the trees. She found that the orangutans used an upright gait, very human-like, to reach the most distal branches where the ripe fruit was [77].

They used the same walking technique when they moved from the crown of one tree to another. They often used one hand on the branches above them for support, but the important observation is that they stretched their legs and stood upright in a way that is quite uncommon among apes. Perhaps the upright gait in the trees was an adaptation for life in the periphery of the tree crowns that offered good protection from predators and where the food was to be found.

Homo sapiens: The Wise Man

According to the classic perception our own species, *Homo sapiens*, appeared in its anatomically original form about 200,000 years ago. Most probably the areas of origin of modern man were the coastline around South Africa and the flatland around the Cape. A small group of early humans may have become isolated here at a time when Africa underwent very profound climate changes, and when the survival of individuals and the continuous development of man may have, for a short period, been in great danger, in fact our ancestors might have been very close to extinction. Easy access to marine food was probably a key issue, and shellfish which are rich in proteins, as well as mussels, limpets and sea snails are thought to have aided survival of the local population. Probably just a few thousand individuals survived and constituted the basis for the continuous development and growth of mankind. The modern characteristics of man might have arisen in such an isolated environment [40, 78, 79].

However, the earliest fossil bone finds of *Homo sapiens* are spread over the whole African continent: over South Africa (Florisbad), Tanzania (Ngaloba), Kenya (Guomde) and Ethiopia (Omo Kibish) to Morocco (Jebel Irhoud). A fossil named Omo from Ethiopia is perhaps the oldest known find of a human with the same appearance as us. The age of this fossil has been estimated at 195,000 years.

However, Blombos cave in South Africa is where the oldest indications of capacity for symbolic thinking among early humans have been found. Here scientists found pieces of ochre with evident carvings characterised by lines and crosses, a symbolic pattern of unknown significance. Probably the individuals who made these inscriptions were living in small spread-out groups of 100–200 individuals. The ability to think in symbols is important for creating and maintaining social networks and indicates individuals with a good intellect. Symbolic thinking may be associated with a belief in a spiritual world and may have improved the possibilities for cooperation and interaction required for survival in a hard reality. Modern research shows that fruitful cooperation among individuals is an innate human activity that stimulates the reward centres of the brain, thereby constituting an important survival mechanism. The capacity for symbolic thinking is closely linked to language, dancing and singing. Probably empathy and a capacity for symbolic thinking were prerequisites for establishing and maintaining group formations that were strong enough to secure the survival and spread of man over the earth.

Out of Africa

How *Homo sapiens* once spread over the earth is a deeply fascinating story [80, 81]. Today we know that all human beings outside Africa originated from a very small group of African ancestors – perhaps a few thousand individuals – who lived a couple of hundred thousand years ago somewhere on the African continent. Individuals from this group of *Homo sapiens* left Africa much later and spread across the earth.

This late emigration was preceded by several earlier emigration waves that had not been as successful. It is believed that the predecessor of *Homo sapiens*, *Homo erectus*, emigrated in several waves towards Asia, probably beginning about 2 million years ago. Java Man and Peking Man, both belonging to the species *Homo erectus*, were descendants of individuals who once left northeast Africa and continued migrating towards the heart of Asia. An early migration also took place towards Europe. *Homo heidelbergensis*, a common ancestor to *Homo neanderthalensis* and *Homo sapiens*, probably emigrated to southern Europe from North Africa in several periods.

Thus, these early emigrants are not our ancestors according to the 'out of Africa theory', originally proposed by Christopher Stringer and Peter Andrews more than 20 years ago [82, 83]. Our own species, *Homo sapiens*, was born in northeast Africa 200,000–170,000 years ago and remained there for an extended time period. But then a new period of emigrations from the African continent began, which over time led to the spread of *Homo sapiens* over the whole earth. Our knowledge about how this happened has increased substantially thanks to a brand-new research instrument that has evolved over the past 40 years – modern DNA technology [84–87].

Much of the new knowledge of human prehistory is based on the 'genetic watch', implying that mutations in the human genome take place at a certain rate. However, it has recently been argued that the mutation rate is considerably slower than was previously believed [88–90]. If this proves to be true, many of the key events in human evolution might have taken place earlier than was previously believed. For instance, our own species, *Homo sapiens*, might have already emerged 250,000–300,000 years ago.

Most likely *Homo sapiens* made several failed attempts to leave the African continent. It has been proposed that one such attempt took place about 120,000 years ago, but these individuals did not get any further than an area corresponding to modern-day Israel. Probably the last major emigration of modern man out of Africa occurred 50,000–60,000 years ago. However, considering the new revised estimation of mutation rate in the 'genetic watch', this exodus out of Africa might have happened much earlier, probably around 100,000 years ago, at a time corresponding to the first finds of emigrants in Israel. These individuals, perhaps just a few groups of a couple of hundred individuals, passed eastwards south of the Red Sea. They travelled east along the shores of the Indian Ocean towards India. Spreading at a rate of about 20 km per generation, it took less than 10,000 years to reach Southeast Asia and the Indonesian archipelago. In this way man spread over all parts of the world, replacing *Homo erectus* in Asia and *Homo neanderthalensis* in Europe, who had emigrated at a much earlier time. In Europe *Homo sapiens* and *Homo neanderthalensis* lived alongside one another for a period of about 15,000 years. Recently published finds, indicating that our genome contains traces of the Neanderthal genome [48, 91], make it necessary to re-evaluate some older concepts: we have to reappraise the established opinion that modern man has an absolute unique genome without interference from other species. *Homo sapiens*, emigrating from Africa at a late time period, have intermingled with *Homo neanderthalensis*, descendants of earlier emigrants from Africa – a concept that was has been impossible to imagine earlier.

Modern man reached Australia about 45,000 years ago. One group headed towards Europe at an earlier stage and arrived there about 40,000 years ago. An Asian group passed the then-dry Bering Strait, reaching the west coast of North America about 20,000 years ago and the coast of South America about 10,000 years later. At about 10,000 B.C., these emigrants reached Scandinavia and southern Sweden at a time when the ice had receded sufficiently to make life possible in this area.

References

1. Johanson DC, Edey MA. Lucy: the beginnings of humankind. London: Penguin; 1981.
2. White TD, Asfaw B, Beyene Y, Haile-Selassie Y, Lovejoy CO, Suwa G, et al. Ardipithecus ramidus and the paleobiology of early hominids. Science. 2009;326(5949):75–86.
3. Lovejoy CO, Simpson SW, White TD, Asfaw B, Suwa G. Careful climbing in the Miocene: the forelimbs of Ardipithecus ramidus and humans are primitive. Science. 2009;326(5949): 70e1–8.
4. Navarrete A, van Schaik CP, Isler K. Energetics and the evolution of human brain size. Nature. 2011;480(7375):91–3.
5. Psouni E, Janke A, Garwicz M. Impact of carnivory on human development and evolution revealed by a new unifying model of weaning in mammals. PLoS One. 2012;7(4):e32452.
6. Napier JR, Tuttle RH. Hands. Princeton: Princeton University Press; 1993.
7. Wilson FR. The hand: how its use shapes the brain, language, and human culture. 1st ed. New York: Pantheon Books; 1998.

8. Gärdenfors P. How homo became sapiens: on the evolution of thinking. New York: Oxford University Press; 2006.
9. Corballis MC. From hand to mouth: the origins of language. Princeton: Princeton University Press; 2002.
10. Mithen SJ. The singing Neanderthals: the origins of music, language, mind and body. London: Weidenfeld & Nicolson; 2005.
11. Tattersall I. Masters of the planet: the search for our human origins. New York: Palgrave Macmillan; 2012.
12. Scientific American. What makes us human. Scientific American, Special collector's edition. 2013;22(1).
13. Leakey MD, Hay RL. Pliocene footprints in the laetolil beds at laetoli, northern Tanzania. Nature. 1979;278:317–23.
14. Hay RL, Leakey MD. Fossil footprints in laetoli. Sci Am. 1982;246:38–45.
15. Agnew N, Demas M, Leakey MD. The Laetoli footprints. Science. 1996;271(5256):1651–2.
16. Green DJ, Alemseged Z. Australopithecus afarensis scapular ontogeny, function, and the role of climbing in human evolution. Science. 2012;338(6106):514–7.
17. Bush ME, Lovejoy CO, Johanson DC, Coppens Y. Hominid carpal, metacarpal, and phalangeal bones recovered from the Hadar formation: 1974–1977 collections. Am J Phys Anthropol. 1982;57:651–77.
18. Marzke MW, Pouydebat E. Comments on E. Pouydebat, P. Gorce, Y. Coppens, V. Bels. Biomechanical study of grasping according to the volume of the object: human versus non-human primates. J Biomech. 2009;42:266–72. J Biomech. 2009;42(15):2628–9.
19. Marzke MW. Origin of the human hand. Am J Phys Anthropol. 1971;34(1):61–84.
20. Marzke MW. Upper-limb evolution and development. J Bone Joint Surg Am. 2009;91 Suppl 4:26–30.
21. Almquist EE. Evolution of the distal radioulnar joint. Clin Orthop Relat Res. 1992;275:5–13.
22. McPherron SP, Alemseged Z, Marean CW, Wynn JG, Reed D, Geraads D, et al. Evidence for stone-tool-assisted consumption of animal tissues before 3.39 million years ago at Dikika, Ethiopia. Nature. 2010;466(7308):857–60.
23. Alemseged Z, Spoor F, Kimbel WH, Bobe R, Geraads D, Reed D, et al. A juvenile early hominin skeleton from Dikika, Ethiopia. Nature. 2006;443(7109):296–301.
24. Wong K. Lucy's baby. Scientific American, Special collector's edition. 2013;22(1):4–11.
25. Haile-Selassie Y, Saylor BZ, Deino A, Levin NE, Alene M, Latimer BM. A new hominin foot from Ethiopia shows multiple Pliocene bipedal adaptations. Nature. 2012;483(7391):565–9.
26. Lovejoy CO, Suwa G, Simpson SW, Matternes JH, White TD. The great divides: Ardipithecus ramidus reveals the postcrania of our last common ancestors with African apes. Science. 2009;326(5949):100–6.
27. Suwa G, Asfaw B, Kono RT, Kubo D, Lovejoy CO, White TD. The Ardipithecus ramidus skull and its implications for hominid origins. Science. 2009;326(5949):68e1–7.
28. Lovejoy CO, Suwa G, Spurlock L, Asfaw B, White TD. The pelvis and femur of Ardipithecus ramidus: the emergence of upright walking. Science. 2009;326(5949):71e1–6.
29. Lovejoy CO, Latimer B, Suwa G, Asfaw B, White TD. Combining prehension and propulsion: the foot of Ardipithecus ramidus. Science. 2009;326(5949):72e1–8.
30. Gibbons A. Breakthrough of the year. Ardipithecus ramidus. Science. 2009;326(5960):1598–9.
31. Leakey MD. Olduvai Gorge: my search for early man. London: Collins; 1979.
32. Leakey LS, Tobias PV, Napier JR. A New species of the genus Homo from Olduvai Gorge. Nature. 1964;202:7–9.
33. Kemble JV. Man's hand in evolution. J Hand Surg Br. 1987;12(3):396–9.
34. Marzke MW. Man's hand in evolution. J Hand Surg Br. 1988;13(2):229–30.
35. Linscheid RL. The hand and evolution. J Hand Surg Am. 1993;18(2):181–94.
36. Bennett MR, Harris JW, Richmond BG, Braun DR, Mbua E, Kiura P, et al. Early hominin foot morphology based on 1.5-million-year-old footprints from Ileret, Kenya. Science. 2009;323(5918):1197–201.

37. Adler R. Fossil footprints reveal our modern walk in the making. New Scientist. 2009; 201(2698):10.
38. Scott GR, Gibert L. The oldest hand-axes in Europe. Nature. 2009;461(7260):82–5.
39. Shen G, Gao X, Gao B, Granger DE. Age of Zhoukoudian Homo erectus determined with (26) Al/(10)Be burial dating. Nature. 2009;458(7235):198–200.
40. Brown KS, Marean CW, Herries AI, Jacobs Z, Tribolo C, Braun D, et al. Fire as an engineering tool of early modern humans. Science. 2009;325(5942):859–62.
41. Leakey MG, Spoor F, Dean MC, Feibel CS, Anton SC, Kiarie C, et al. New fossils from Koobi Fora in northern Kenya confirm taxonomic diversity in early Homo. Nature. 2012;488(7410): 201–4.
42. Pennisi E. Neandertal genomics. Tales of a prehistoric human genome. Science. 2009; 323(5916):866–71.
43. Hall S. Last of the Neanderthals. National Geographics. 1 October 2008(.p. 34–56.
44. Wynn T, Coolidge F. How to think like a Neanderthal. Oxford: Oxford University Press; 2011.
45. Briggs AW, Good JM, Green RE, Krause J, Maricic T, Stenzel U, et al. Targeted retrieval and analysis of five Neandertal mtDNA genomes. Science. 2009;325(5938):318–21.
46. Pennisi E, Ancient DNA. Sequencing Neandertal mitochondrial genomes by the half-dozen. Science. 2009;325(5938):252.
47. Burbano HA, Hodges E, Green RE, Briggs AW, Krause J, Meyer M, et al. Targeted investigation of the Neandertal genome by array-based sequence capture. Science. 2010;328(5979): 723–5.
48. Green RE, Krause J, Briggs AW, Maricic T, Stenzel U, Kircher M, et al. A draft sequence of the Neandertal genome. Science. 2010;328(5979):710–22.
49. Musgrave JH. How dextrous was Neanderthal man? Nature. 1971;233(5321):538–41.
50. Musgrave JH. The Neandertals from Krapina, northern Yugoslavia: an inventory of the hand bones. Z Morphol Anthropol. 1977;68(2):150–71.
51. Niewoehner WA, Bergstrom A, Eichele D, Zuroff M, Clark JT. Digital analysis: manual dexterity in Neanderthals. Nature. 2003;422(6930):395.
52. Wilkins J, Schoville BJ, Brown KS, Chazan M. Evidence for early hafted hunting technology. Science. 2012;338(6109):942–6.
53. Zilhao J, Angelucci DE, Badal-Garcia E, d'Errico F, Daniel F, Dayet L, et al. Symbolic use of marine shells and mineral pigments by Iberian Neandertals. Proc Natl Acad Sci U S A. 2010;107(3):1023–8.
54. Balter M. Archaeology. Neandertal jewelry shows their symbolic smarts. Science. 2010; 327(5963):255–6.
55. Finlayson C. The human who went extinct: why Neanderthals died out and we survived. Oxford: Oxford University Press; 2009.
56. Balter M. Archaeology. Better homes and hearths, Neandertal-style. Science. 2009;326(5956): 1056–7.
57. Brown P, Sutikna T, Morwood MJ, Soejono RP, Jatmiko, Saptomo EW, et al. A new small-bodied hominin from the Late Pleistocene of Flores, Indonesia. Nature. 2004; 431(7012):1055–61.
58. Jungers WL, Harcourt-Smith WE, Wunderlich RE, Tocheri MW, Larson SG, Sutikna T, et al. The foot of Homo floresiensis. Nature. 2009;459(7243):81–4.
59. Weston EM, Lister AM. Insular dwarfism in hippos and a model for brain size reduction in Homo floresiensis. Nature. 2009;459(7243):85–8.
60. Lieberman DE. Palaeoanthropology: Homo floresiensis from head to toe. Nature. 2009; 459(7243):41–2.
61. Brumm A, Jensen GM, van den Bergh GD, Morwood MJ, Kurniawan I, Aziz F, et al. Hominins on Flores, Indonesia, by one million years ago. Nature. 2010;464(7289):748–52.
62. Falk D. The fossil chronicles: how two controversial discoveries changed our view of human evolution. Berkeley: University of California Press; 2011.
63. Tocheri MW, Orr CM, Larson SG, Sutikna T, Jatmiko, Saptomo EW, et al. The primitive wrist of Homo floresiensis and its implications for hominin evolution. Science. 2007;317:1743–5.

64. Wong K. Rethinking the hobbits of Indonesia. Sci Am. 2009;301(5):66–73.
65. Krause J, Fu Q, Good JM, Viola B, Shunkov MV, Derevianko AP, et al. The complete mitochondrial DNA genome of an unknown hominin from southern Siberia. Nature. 2010;464(7290):894–7.
66. Brown TA. Human evolution: stranger from Siberia. Nature. 2010;464(7290):838–9.
67. Dalton R. Fossil finger points to new human species. Nature. 2010;464(7288):472–3.
68. Berger LR, de Ruiter DJ, Churchill SE, Schmid P, Carlson KJ, Dirks PH, et al. Australopithecus sediba: a new species of Homo-like australopith from South Africa. Science. 2010;328(5975): 195–204.
69. Dirks PH, Kibii JM, Kuhn BF, Steininger C, Churchill SE, Kramers JD, et al. Geological setting and age of Australopithecus sediba from southern Africa. Science. 2010;328(5975): 205–8.
70. Pickering R, Dirks PH, Jinnah Z, de Ruiter DJ, Churchil SE, Herries AI, et al. Australopithecus sediba at 1.977 Ma and implications for the origins of the genus Homo. Science. 2011; 333(6048):1421–3.
71. Lawton G. Could the orangutan be our closest relative? New Scientist. 2009;202(2713):6–7.
72. Schmitt D. Insights into the evolution of human bipedalism from experimental studies of humans and other primates. J Exp Biol. 2003;206(Pt 9):1437–48.
73. Richmond BG, Begun DR, Strait DS. Origin of human bipedalism: the knuckle-walking hypothesis revisited. Am J Phys Anthropol. 2001;Suppl 33:70–105.
74. Senut B, Pickford M, Gommery D, Mein P, Cheboi K, Coppens Y. First hominid from the Miocene (Lukeino Formation, Kenya). C R Acad Sci Paris, Sciences de la Terre et des planètes. 2001;332:137–44.
75. Morgan E. The scars of evolution. New York: Oxford University Press; 1994.
76. Morgan E. The aquatic ape hypothesis. London: Souvenir; 1999.
77. Thorpe SK, Holder RL, Crompton RH. Origin of human bipedalism as an adaptation for locomotion on flexible branches. Science. 2007;316(5829):1328–31.
78. Henshilwood CS, Marean CW. The origin of modern human behavior. Curr Anthropol. 2003;44(5):627–51.
79. Marean CW, Bar-Matthews M, Bernatchez J, Fisher E, Goldberg P, Herries AI, et al. Early human use of marine resources and pigment in South Africa during the Middle Pleistocene. Nature. 2007;449(7164):905–8.
80. Stringer C. Human evolution: Out of Ethiopia. Nature. 2003;423(6941):692–3, 5.
81. Clark JD, Beyene Y, WoldeGabriel G, Hart WK, Renne PR, Gilbert H, et al. Stratigraphic, chronological and behavioural contexts of Pleistocene Homo sapiens from Middle Awash, Ethiopia. Nature. 2003;423(6941):747–52.
82. Stringer C. Modern human origins: progress and prospects. Philos Trans R Soc Lond B Biol Sci. 2002;357(1420):563–79.
83. Klein RG. Out of Africa and the evolution of human behaviour. Evol Anthropol. 2008; 17:267–81.
84. Stix G. Traces of a distant past. Sci Am. 2008;299(1):56–63.
85. Wade N. Before the dawn: recovering the lost history of our ancestors. New York: Penguin Press; 2006.
86. Desalle R, Tattersall I. Human origins: what bones and genomes tell us about ourselves. College Station: Texas A & M University Press; 2008.
87. Weaver DT, Roseman CC. New developments in the genetic evidence for modern human origins. Evolutionary Anthropology. 2008;17:69–80.
88. Scally A, Durbin R. Revising the human mutation rate: implications for understanding human evolution. Nat Rev Genet. 2012;13(10):745–53.
89. Langergraber KE, Prufer K, Rowney C, Boesch C, Crockford C, Fawcett K, et al. Generation times in wild chimpanzees and gorillas suggest earlier divergence times in great ape and human evolution. Proc Natl Acad Sci U S A. 2012;109(39):15716–21.
90. Callaway E. Studies slow the human DNA clock. Nature. 2012;489(7416):343–4.
91. Hofreiter M. Drafting human ancestry: what does the Neanderthal genome tell us about hominid evolution? Commentary on Green et al. (2010). Hum Biol. 2011;83(1):1–11.

Chapter 3
The Hand, the Brain and Tools

Abstract Three to four million years ago, the early hominins had well-developed hands, but their brains were too rudimentary for tool making. Even if Ardi (*Ardipithecus ramidus*) and Lucy (*Australopithecus afarensis*) were able to use stones and other objects to protect themselves or attack enemies, there is no indication that they were able to make tools by themselves. Tool making requires imagination and a capacity for planning and abstract thinking, and it has been proposed that the larger *Homo habilis* brain was required for tool making to become possible. However, new findings in Ethiopia's Afar region indicate that tool making might have taken place about 800,000 years earlier than previously believed. A large brain may not be a prerequisite for tool making after all, as indicated by the advanced tools made by 'the hobbit', *Homo floresiensis*, as well as various types of tools used by chimpanzees, crows and rooks. Some researchers now suggest that man's use of tools may have been a contributing factor to the development of the brain rather than vice versa and that the use of technical aids might have increased the chances of survival and thus had an essential influence on the development of humankind and societal structures.

The very early hominins, living 3–4 million years ago, had well-developed hands, but their brains were too rudimentary for tool making. Even if Ardi and Lucy were able to use stones and other nearby objects for various purposes, so far there have been no indications that they were able to make tools themselves. Tool making requires imagination and a capacity for planning and abstract thinking, and it has been proposed that tool making first became possible due to the larger *Homo habilis* brain. But new findings, published in *Nature* in 2010, opened new perspectives on the issue of early tool making. Zeresenay Alemseged and associates from the California Academy of Sciences have been conducting 'the Dikika Research Project' in Ethiopia's Afar region for many years. This is where fossils of hominins like Lucy were found. Marks and damage on fossilised antelope bones, dated 3–4 million years ago, clearly show that hominins used stone tools to scrape meat from bones and gain access to the bone marrow [1], about 800,000 years earlier than previously believed.

Ardi and Lucy could probably pick up and throw stones at attacking enemies, but no one can say for sure if they could also make more permanent tools for specific purposes, such as hunting and defence. It was believed that only *Homo habilis* presented such characteristics, and Louis Leakey argued strongly that *Homo habilis* was the first prehuman with a sufficiently large brain to be able to construct and make tools. However, these well-established theories are now being called into question by the data presented by Alemseged's team.

Homo erectus has a larger brain, and the tool making associated with this species, called the Acheulian culture, is much more advanced. The tools included hand axes shaped by striking flakes off both sides of a stone to create a sharp edge. A much later type of tool making, the Mousterian culture, dated 200,000–100,000 years ago, is associated with *Homo sapiens* and represents a much more refined design than earlier tools. This was understood to be an indication of a more advanced capacity for imagination, planning, design and abstract and organised thinking, abilities that characterise a more developed brain. More complicated tools were made in which sharpened stones were mounted on shafts and handles to make spears and axes, projectiles and harpoons. Decorative items, including various types of ornaments, became common.

Is a larger brain really a prerequisite for making tools? Certainly not. Signs of advanced tool making were also demonstrated by *Homo floresiensis*, our newly discovered 'cousin' who lived on isolated islands in Southeast Asia until 13,000 years ago. These individuals, about 1 m in height, had extremely small brains with a volume of about 400 cc [2] but well-developed frontal lobes in the brain, indicating higher brain functions and a capacity for imagination, planning and abstract thinking [3]. Since 2002, the archaeologist Mark Moore at New England University in Armidale, Australia, has been studying stone tools made by *Homo floresiensis*, which were discovered in the Liang Bua cave on the island of Flores [4]. Tools made by *Homo floresiensis* were found in different layers of the cave along with tools originating from *Homo sapiens*. Some striking similarities in working techniques have been identified, and it has been speculated that these two species may at one time have lived side by side and that individuals belonging to *Homo sapiens* even learned from their cousins, *Homo floresiensis* [4]. So it makes no sense to consider only the size of the brain when it comes to a capacity for tool making; what is important, of course, is what the brain is capable *of doing*. The inner organisation of the brain is not reflected in its size, and indeed, the fossil findings say nothing about the inner organisation of these brains and their synaptic networks.

There are several examples of how animals with small brains can not only use tools but also, in a very clever way, determine how to make them [5]. Among birds, finches use spines from vegetation to search for insects in tree cracks, and thrushes as well as sea otters use stones to crush the shells of mussels and clams. There are stories about rooks dropping nuts on roads with heavy traffic so that cars will crush their shells; very smart individuals have been observed dropping the nuts on pedestrian crossings and then waiting for the 'green light' when the traffic stops and they can easily pick up the readily prepared food without the risk of being run over by cars. During antiquity, there were descriptions of how intelligent crows and rooks could drop stones into narrow containers of water to raise the water level so the bird could reach a delicious caterpillar on the surface [6].

It has been argued that only modern man and some of his early ancestors have the capacity to *make* tools. But rooks are known to be able to pick up edible titbits caught in narrow spaces after having intentionally bent the tip of a steel wire to make a hook [6]. In the 1980s, Jane Goodall documented how chimpanzees can very carefully prepare and use tools and how they teach youngsters the art of tool making. She observed how chimpanzees stripped leaves from branches to 'fish' for termites. Chimpanzees also use chewed-up leaves as sponges to absorb water out of crevices.

Chimpanzees are also capable of planning how to make tools as well as the future use of tools or weapons. In 2009 Mathias Osvath from Lund University in Sweden described how a stone-throwing chimpanzee in the Furuvik Animal Park in Skutskär prepared attacks on the park visitors. Every morning, the male chimpanzee systematically collected stones and concrete fragments that he later angrily threw at the visitors [7].

It is believed that the planning, making and use of tools greatly influenced the development of our brain. A very old bone fragment from a gnu in Ethiopia shows indentations that indicate that the bone was notched with stone tools 2.5 million years ago – half a million years before *Homo erectus* with its fairly well-developed brain appeared. The finds support the theory that we did not start making tools because we had big brains, but rather that the use and making of tools were important factors behind the development of our brains and our intelligence.

But might it perhaps have been the other way around – that other factors stimulated the development of the brain, resulting in an increased ability to use the capacities of the hand to make tools? Africa has undergone several periods of very extensive climatic changes, which also influenced man's relationship with nature. These climate changes may have indirectly contributed to the development of modern man and his big brain. For example, groups of individuals might have assembled at the seashores and become used to marine food such as fish, mussels, clams and crustaceans. This kind of food is rich in proteins and important fatty acids that may have contributed to the development of a big brain. There is a direct relationship between access to food and brain development, and it is well known that omega-3 fatty acid is one of several nutritional substances that benefit and support the evolution of the brain. Docosahexaenoic acid (DHA) is one of the most common omega-3 fatty acids in the cell membranes of the brain and is highly present in fish and other kinds of seafood.

Perhaps the combination of improved nutritional intake and more advanced use of the hand were both important beneficial factors supporting brain development. The upright posture that freed up our hands, and associated changes in the hands' functions, probably resulted in an adaptation process in the brain cortex, reorganising neural networks, but such a process requires a substrate: cellular and matrix components that make it possible for neural networks to expand and the brain matter to grow. It is well known that functional reorganisations in the brain cortex are stimulated by an enriched environment, challenges and processes involving problem-solving. Perhaps enlargement of the brain occurred primarily during periods of new living conditions in environments where new ways of thinking and innovative processes were necessary for survival.

While the possibilities created by free hands may have stimulated development of the brain, simply walking upright may have been important for the growth of bigger brains. The two-footed gait may have required greater balancing efforts, which in turn required the sensory organs to interact more intensely – especially vision, hearing and the sensitivity of the feet. The fact that the arms and hands could operate independently of the lower extremities demanded even more brain power and control.

Homo Handmade

The deceased architect, humanist and society builder Peter Broberg created the expression *technolution* as a descriptive term for man's biological evolution. Broberg agrees with the opinion that the hand has played a major role in the evolution of man, and he posits that the use of *technology* – tools, weapons – had an essential influence on man's physical appearance and development. He says that early use of technology among hominins led to more complex behaviour, which in itself increased the need for more brain capacity. He also believes that language, which is also needed in a more complex environment, probably appeared at the same time. According to this conceptual use of technology, language and brain development occurred in parallel and probably influenced each other. According to Broberg man developed from the early hominin stage through technological advancements which continuously contributed to the extension and improvement of the capacity of the human body. Thus, technology created man rather than the other way around – 'technology is the father of man and language is his mother'. Broberg regards man as a *Homo sapiens technicus*, and considering the central role of the hand in the development of the final product, he would prefer the name *Homo handmade* as an alternative to *Homo sapiens*.

Early in evolution the hand, the brain and the tools we used interacted in a way that promoted the evolution of societal structure. Early tools became the basis for developing improved weapons and hunting technology, but in addition to this, a new abstract thinking was required – the ability to plan, to design and to resolve the challenges of using and improving tools. And so tools developed man just as man developed tools. Thomas Johannesson, former dean at Lund Technical College, describes how the development has been characterised by the interaction between man's inner activities (thoughts) and his outer activities (handicrafts). New concepts and thoughts led to new tools that in turn inspired new thoughts, a pattern that is evident in technology as well as science and art.

Technology is about improving, enhancing and simplifying, and here the hand has a central role. A prerequisite for using technology is the hand's gripping function, rotational movements in the forearm and the sensitivity of the hand. Tools can be regarded as extensions of the hand, actually changing the natural limitations of the hands' abilities. The technology-enhanced individual will enjoy greater capacity and freedom. Examples of an important 'body extension' are the first weapons, which improved the chances of survival. The first settlements were associated with several new technologies – hunted quarry had to be transported to the dwelling place, the food had to be prepared and stored and the children had to be protected. Migration to the northern hemisphere required improved clothing, the use of open fire and new types of dwellings.

The mechanisation of agriculture is a good example of how man has enhanced and amplified the capacity of the hand through machines and more effective tools. The introduction of the first machines in the textile industry heralded a new era where the capacity of the hand could be multiplied several times, creating a new dimension of complexity and effectiveness.

One example of the early use of technology is using a sewing needle and vegetable fibres to make strings, baskets and clothes. Fragments of linen fibres, which may have been used for such purposes 30,000 years ago, have been found in the Dzudzuana cave in Georgia [8]. The fibres may have been important for fixating and combining pieces of stone tools. The finds also included spun fibres, sometimes with several knots. Black, grey and turquoise fibres were also found, indicating that the inhabitants of the cave could make dyed textiles.

Another example of the importance of technical achievements is the use of bags and knapsacks. The cupped hand was the first container that could bring water to the mouth and could hold and transport small amounts of materials. Later, it was supported by objects made of animal skin or woven plant fibres, allowing people to collect and transport nuts, fruits, roots and small quarry back to the settlement. Thus, the bag became important for cooperation between individuals and social organisation as it became possible to carry home food and share it with other members of the group. The San people in the Kalahari Desert made good use of bags made of carefully prepared antelope skin, which were carried on the shoulder or by a leather strap around the waist. A bag also played a major role for those who wanted to carry a baby while gathering food. Woven baskets may also have played an important role for the ability to carry food and various objects during migrations over long distances.

References

1. McPherron SP, Alemseged Z, Marean CW, Wynn JG, Reed D, Geraads D, et al. Evidence for stone-tool-assisted consumption of animal tissues before 3.39 million years ago at Dikika, Ethiopia. Nature. 2010;466(7308):857–60.
2. Weston EM, Lister AM. Insular dwarfism in hippos and a model for brain size reduction in Homo floresiensis. Nature. 2009;459(7243):85–8.
3. Falk D, Hildebolt C, Smith K, Morwood MJ, Sutikna T, Brown P, et al. The brain of LB1. Homo floresiensis. Science. 2005;308(5719):242–5.
4. Culotta E. Archaeology. Did humans learn from hobbits? Science. 2009;324(5926):447.

5. Haslam M, Hernandez-Aguilar A, Ling V, Carvalho S, de la Torre I, DeStefano A, et al. Primate archaeology. Nature. 2009;460(7253):339–44.
6. Marzluff J, Angell T. Gifts of the crow: how perception, emotion, and thought allow smart birds to behave like humans. New York: Free Press; 2012.
7. Osvath M. Spontaneous planning for future stone throwing by a male chimpanzee. Curr Biol. 2009;19(5):R190–1.
8. Kvavadze E, Bar-Yosef O, Belfer-Cohen A, Boaretto E, Jakeli N, Matskevich Z, et al. 30,000-year-old wild flax fibers. Science. 2009;325(5946):1359.

Chapter 4
How the Hand Generated Language

Abstract Hands and arms are important components of communication and inter-action among individuals. From an evolutionary perspective, bipedalism made it possible to use the hands for communication by gestures and signs long before there was a spoken language. Speech as well as hand/arm movements are associated with the activation of Broca's area in the left frontal cortex, and it appears that Broca's area is involved in the organisation of speech as well as hand movements. In our normal body language and our communication with other individuals, gestures and hand movements are linked to speech in a natural way, reflecting overlapping repre-sentational areas of these functions in Broca's area. The gene FOXP2, necessary for coordination of tongue and lip movements in a spoken language, seems largely to have appeared in the last 200,000 years of human development. Thus, the condi-tions necessary for fully developed speech seem to be have appeared late in the evolutionary process, about the time our own species, *Homo sapiens*, emerged.

A unique human trait is the ability to communicate using a spoken language, creat-ing a basis for our social life and giving our culture a special dimension. When we speak we can formulate, express and convey thoughts and ideas in our communica-tion with others. The benefits of a spoken language are enormous. Sound has few limitations and can be transmitted around corners and over large distances. It is not suppressed by darkness, and it does not require eye contact with the person being addressed. The ability to articulate and to vary the volume and pitch of the voice introduced great potential for high precision in verbal communication.

The arms and hands also play an important role in communicating with other individuals. Signs and gestures are obvious and natural complements to the spoken language. It seems that talking and hand/arm movements are integrated and insepa-rable components in our communication with others.

How did our earliest ancestors communicate with each other? How and when during evolution did the spoken articulated language emerge? Today we know that there are several requirements for making an articulated vocal language possible. The brain must be sufficiently well developed for abstract and symbolic thinking to occur, which is necessary to make a spoken language possible. Appropriate anatomical conditions must be there to make possible the production of precise and meaningful sounds. Also, the larynx and the vocal cords must have the right anatomical structures and be positioned correctly, and they must have a sufficient nerve supply. In addition, there must be an ability to control breathing with rapid inhaling and slow exhaling, making the vocal cords vibrate appropriately.

Such conditions were first present late in the evolutionary process. So how did our early ancestors communicate with each other? Probably by using gestures, facial expressions and simple sounds, perhaps unarticulated grunts. There are strong indications that human language in fact emerged from body language, especially from gestures of the hands and facial expressions, and that the first language in fact was a language of gestures – a primitive sign language.

In his book *From Hand to Mouth*, Corballis argues that human language was initially based on gestures rather than sounds and that bipedalism and upright walking were important steps in making a language of gestures possible – a signed language where the free arms and hands allowed gestures, signs and hand movements [1]. A language based on gestures and signs is in fact very rich in expressions; yet it is simple and does not require complicated anatomical structures for sound generation.

A signed language can have many advantages: one can use it to avoid detection – gestures are silent while speaking is easy for enemies to hear. The hand can easily indicate directions and objects and point out phenomena in the environment. In addition, it is easy to indicate the size and shape of objects using the hands. A newborn child points before it is able to speak. All of us gesticulate while we are talking; the hand is often described as an important organ for communication. The modern sign language of the deaf and people with impaired hearing is rich and complete and can, in some respects, have several advantages over spoken language [2].

Spoken language and the language of gestures have a close functional and structural connection: sign language and spoken language activate the same area in the brain, Broca's area, situated in the left frontal cortex, just in front of the area that controls the movements of the mouth and hands. Broca's area is critically involved in the organisation of spoken language. It is not always easy to separate movements in the hand from movements in the mouth and tongue; fine motor movements in the hand during intense concentration are also often associated with movements of the tongue and mouth.

Broca's area is also activated in deaf people using their hands and arms in sign language – the same area in the brain that is used for a spoken language in hearing people [3, 4]. Broca's area is rich in *mirror neurons*, nerve cells that are activated when hand/arm movements are performed and when corresponding movements performed by somebody else are observed (see also Chap. 11). Thus, the mirror neurons form a basis for imitation and learning by observation, and it has been supposed that they therefore have an important role in learning a language of gestures and, indirectly, spoken language [5]. Observation of gestures and hand movements activates the observer's own motor programs for corresponding movements, and the muscles regulating tongue movements and speech become involved in the process.

There is probably a close link between language and making tools. The relationship between the body's motor functions, language and thought is well documented, and the development of language is intimately linked to the body's ability to express a body language. Making tools requires a well-developed intellect, and transferring knowledge and experience from tool making probably requires some kind of speaking ability.

So, when during evolution did the language of the hand become a spoken language? When can we find a hint of an area in the brain that provides for an articulated speaking ability? It has been possible to make 'endocasts' of the brain by filling fossil skull cavities with sediment. Casts of Broca's area have been made from early forms of *Homo* but not from the much older *Australopithecus*. According to the fossil finds, Broca's area began developing about 2.5 million years ago, but obvious imprints from this area have only been found in *Homo erectus*. Thus, it seems that physiological and anatomical conditions in our brains making articulated speech possible began to occur about 2 million years ago.

In addition, articulated speech requires a larynx situated deep in the throat (deeper than among the apes) and wilful control of breathing. Transfer of the larynx to a deeper position may in fact be a result of bipedalism. It requires that the *foramen magnum* – the big hole allowing the spinal cord entry into the skull cavity – is situated on the underside of the skull: the hole is transposed forwards as compared to the situation in four-footed animals, and the skull itself has been transposed backwards so that it can balance on the top of the vertebral column. This requires a smaller jaw, an extension of the vocal tract and a downward transposition of the larynx – a prerequisite for speaking. These changes gradually emerged during evolution, but fossil finds indicate that our ancestor, *Homo erectus*, began acquiring the capacity for articulation about 2 million years ago, something no hominins had earlier.

The tongue receives its nerve supply from the hypoglossal nerve, which passes through a bone canal at the base of the skull. This canal is narrow in apes and *Homo habilis*, but in early *Homo sapiens*, it is widened, a fact that could indicate that *Homo*, in contrast to earlier prehumans, began acquiring traits necessary for articulated speech. Fossil finds indicate that the nerve supply of the tongue may have been fully developed about 300,000 years ago, indicating that a relatively fully developed speaking ability may have been present at that time.

There are calculations indicating that the gene FOXP2 – which is necessary for coordinating the movements of the tongue and lips in a spoken language – was first present in human development over the last 200,000 years [6–9]. Thus, the conditions necessary for fully developed speech seem to have first appeared very late in evolution, about at the time of the appearance of our own species, *Homo sapiens*.

References

1. Corballis MC. From hand to mouth: the origins of language. Princeton: Princeton University Press; 2002.
2. Sacks OW. Seeing voices: a journey into the world of the deaf. Berkeley: University of California Press; 1989.

3. Hickok G, Bellugi U, Klima ES. The basis of the neural organization for language: evidence from sign language aphasia. Rev Neurosci. 1997;8(3–4):205–22.
4. McGuire PK, Robertson D, Thacker A, David AS, Kitson N, Frackowiak RS, et al. Neural correlates of thinking in sign language. Neuroreport. 1997;8(3):695–8.
5. Rizzolatti G, Sinigaglia C. Mirrors in the brain: how our minds share actions and emotions. Oxford: Oxford University Press; 2008.
6. Zhang J, Webb DM, Podlaha O. Accelerated protein evolution and origins of human-specific features: Foxp2 as an example. Genetics. 2002;162(4):1825–35.
7. Vargha-Khadem F, Gadian DG, Copp A, Mishkin M. FOXP2 and the neuroanatomy of speech and language. Nat Rev Neurosci. 2005;6(2):131–8.
8. Dominguez MH, Rakic P. Language evolution: the importance of being human. Nature. 2009;462(7270):169–70.
9. Enard W, Przeworski M, Fisher SE, Lai CS, Wiebe V, Kitano T, et al. Molecular evolution of FOXP2, a gene involved in speech and language. Nature. 2002;418(6900):869–72.

Chapter 5
Handprints from the Past

Abstract Hands are frequently represented in cave and cliff art found in various locations worldwide. Most well known are the handprints found in caves in southern Europe, primarily France and Spain as well as Italy, Portugal, Germany and the Balkans. Handprints are also found outside Europe, for instance in Australia, North America and South America. The most common are so-called negative handprints, where coloured pigment has been sprayed around a hand placed on a cliff wall to produce a 'negative' hand image. In some locations, the images show mutilated hands lacking one or more fingers. The meaning of the hand images is not clear, but they may have been important components of religious rituals. In Sweden hands are frequently depicted in rock carvings in the county of Bohuslän in the southwest part the country. A spectacular rock carving from the late Bronze Age 3,000–3,500 years ago, portraying a man and a woman unified by a large joint hand, symbolises 'the sacred union' between the god and goddess of fertility. This carving, like a large number of additional carvings at the site, is covered by heavy vegetation, making them available only to a few experts in the field.

What is the meaning and symbolism of the paintings and handprints from the past that have been found in caves and on cliffs worldwide? The pictures are deeply fascinating and are usually of very high artistic quality. The paintings are dominated by animal pictures, but images of hands are also common. The locations where cave and cliff art have been discovered reflect our ancestors' migration over the earth after they left the African continent at some time 50,000–60,000 years ago.

Most cave paintings exhibiting animal and hand motifs were discovered in southern Europe, especially in France and Spain [1–3]. Many of them were discovered quite accidentally. When Marcelino Sanz de Sautuola discovered the first paintings in the Altamira cave in Cantabria, Spain, in 1879, the experts of the time believed they were a falsification. Twenty years later it was believed that the paintings in fact represented genuine artistic expressions from our ancestors. Since then almost 350 caves with animal paintings and handprints have been discovered in France and Spain, as well as in Italy, Portugal, Germany and the Balkans [4–16].

G. Lundborg, *The Hand and the Brain*,
DOI 10.1007/978-1-4471-5334-4_5, © Springer-Verlag London 2014

The same types of images were also discovered in North and South America, Africa and Australia.

We do not know why our ancestors produced these images and left them for later generations. Perhaps it was all about a religious ritual, a marking of ownership or perhaps a wish to leave a signature behind for posterity. Several of the handprints are located in remote parts of the caves that were not inhabited and could not be reached without great effort. In some caves, one has to crawl several hundred metres through long, narrow passages to reach the handprints. This indicates that the images probably had some ceremonial or religious meaning. Perhaps by printing his hand on the cliff wall, the shaman could force his way to the other side to reach the world of the gods.

The Hand in the Cave

Hand images were created in several ways. Positive hands are actual prints of hands where the palm was covered with a pigment. This type of handprint is relatively uncommon. More common are negative images where the hand has been pressed against the wall of the cave and the surface around it has been coloured with pigment (Fig. 5.1). It is believed that the pigment was sucked into the mouth and then sprayed over the hand and the surrounding cliff wall so that a negative hand image was created [8, 17, 18]. Some images were probably made by artists holding a pigment-filled tube in one of the hands and blowing through the tube to spray the pigment on the rock surface around the splayed target hand [3]. Probably pigment from red ochre (*hematite*), giving a red colour, and manganese dioxide, giving a black colour, was used. Perhaps the pigment was first dissolved in water to give a better spray effect. In some cases, the pigment seems to have been applied manually to the cliff wall with help of some soft object.

Most negative handprints were discovered in Spain and France. In the Castillo cave in Cantabria, there are 64 handprints of which 55 are left hands, five are right hands and four are inconclusive. The Gargas and Tibiran caves in the French Pyrenees boast more than 60 handprints, and the Cosquer and Chauvet caves in southern France have a large number of negative handprints. Some date to 27,000 years ago, and others might be older.

There are two types of negative handprints: normal hands and defective hands, lacking one or more fingers or parts of fingers (Fig. 5.1). Many theories have been proposed to explain such images exhibiting injured hands. Most likely the hands were purposely mutilated for some ritualistic reason, perhaps in connection with initiation rites. Such mutilated hands are frequent in the Gargas cave, in the nearby Tibiran cave, in the Cosquer cave outside of Marseille and the Pech Merle cave in southwest France [4]. Most often the little finger is amputated at the level of the first middle joint (proximal interphalangeal joint), but there are also several variants. An alternate interpretation of the mutilated hands is that the people of the Ice Age purposely varied the handprints by flexing the fingers to achieve various patterns – perhaps to be able to recognise their own hands.

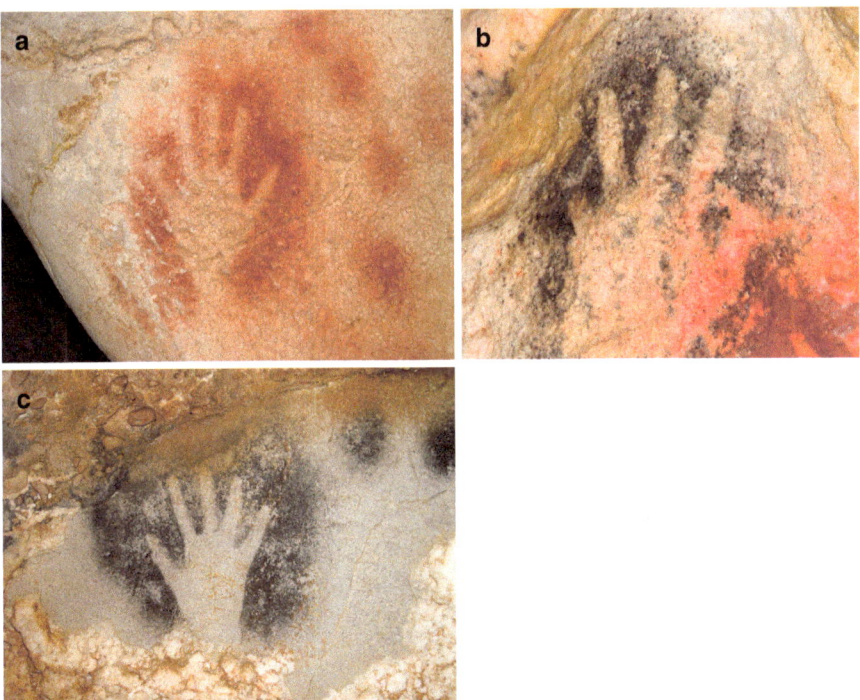

Fig. 5.1 Negative handprints from caves in southern Europe. (**a**) Pigment of red ochre has been sprayed to colour the rock wall around the hand (Pech Merle). (**b**) Hand with the little finger partially amputated (Gargas). (**c**) The rock wall around the hand has been coloured black using mangan powder (Pech Merle) (Photo: Dean Snow)

Left hands are much more common than right hands in cave art. This can probably be explained by the artists' use of the right hand to fill their mouth with pigment to spray the coloured substance around the left hand, placed on the cliff wall. According to Professor Dean Snow at Pennsylvania State University, the size and appearance of the pictures indicate that female hands were depicted in several cases and also children's hands [3].

A large number of negative handprints were also discovered in Patagonia, Argentina. In the Cueva de las Manos (Cave of the Hands) in the Perito Moreno National Park and on the cliffs along the Río Pinturas (the River of Paintings), there are several hundred handprints (Fig. 5.2) combined with hunting scenes and pictures of wild animals, like pumas, which still are present in the area [19]. The age of these handprints has been estimated at 6,000–9,000 years. The images occur both in positive and negative form and in various colours – red, violet, white, black, yellow, orange, ochre and green. In 1999, the Cueva de las Manos and Río Pinturas canyon were declared world heritage sites by UNESCO.

Fig. 5.2 Handprints from the
Rio Pinturas Canyon,
Patagonia, dated 6,000–9,000
years ago (Photo: James
Blair, *National Geographic*,
August 1986)

Hands and Rock Carvings

Sweden has a rich treasure of cliff art in the form of rock carvings. Among the most
famous rock carvings are those discovered in the county of Bohuslän in southwest
Sweden, especially around the village of Tanum and surrounding parishes. The first
documentation of rock carving in Bohuslän was made in the seventeenth century.
Over the last 25 years, almost 3,000 newly discovered rock carvings were registered
and still more continue to be discovered. In several areas, the carvings cover large
areas of bare rock faces that are often covered with moss and therefore are difficult
to discover.

Hands and hand signs are not unusual in the rock carvings of Bohuslän. Lasse
Bengtsson, archaeologist and expert on rock carvings, former chief of the Vitlycke
museum in Tanum, describes several images of ships where arms and hands seem to
grow out from the ship. Often human-like configurations with very big hands are
present in the rock carvings, and Bengtsson tells, with great enthusiasm, about

Fig. 5.3 Drawing of a rock carving from the parish of Askum in Bohuslän in southwestern Sweden. To the *left* in the picture, a man and woman appear unified with a common hand representing *hieros gamos* – the sacred union. On *top* of the man's right hand is a contour of a bear's paw (Courtesy of the Foundation for Documentation of Rock Art in Bohuslän)

several discoveries where the hands have a prominent role in the pictures. We especially discussed an image that has long been used as a logotype for the Scandinavian Society for Surgery of the Hand and represents a man and a woman with one enlarged hand in common. The picture is often used to illustrate the great importance of the hand, not least in communication between individuals. The rock carving represents one small detail in a much larger rock carving in the parish of Askum south of Tanum (Fig. 5.3). The male's right hand is covered by a bear paw. The rock carving was probably produced during the late Bronze Age, approximately 1,000–500 B.C.

Bengtsson explained that the image expresses an important and well-known symbolism and that the picture in fact illustrates the *sacred union* – in Greek *hieros gamos* – of the god and goddess of fertility. To the left in the picture, the goddess of fertility can be seen with spread legs. The god is represented with a phallus and a sword. The hand in common symbolises the unifying of both individuals – the size of the hand probably symbolises a divine force. Their union is completed every autumn, after which the god of fertility disappears to the realm of death for the winter, appearing again in the spring just in time to see the results of his union with the goddess – the rebirth of the world with greenery and flourishing vegetation. A new union takes place the next autumn.

When I visited Lasse Bengtsson in early spring of 2009, he had promised to show me how the rock carvings really look, and he had also promised to make a kind of image, a rubbing, right on the spot of the rock carving. We drove on small, winding roads to a little forest about 40 km south of Tanum, bringing some simple

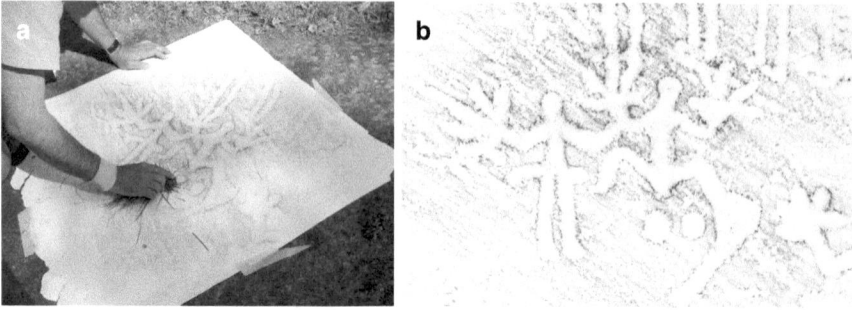

Fig. 5.4 (**a**) Lasse Bengtsson takes a rubbing of the rock carving 'the sacred union'. When he rubs the blueprint paper against the paper, the image – several thousand years old – emerges out of the rock. A magic moment. (**b**) The complete rubbing – the sacred union

tools like a broom, a roll of paper and a roll of blueprint paper. We stopped at a wooded hillside that looked like several other hillsides in the same area; no one would believe that unique rock carvings from the Bronze Age could be hidden in this area. After a 10 min walk, we reached an area with bare hillsides, some overgrown with moss and lichen. Bengtsson started to brush one of the hillsides to remove a big anthill and large amounts of moss. We saw a large number of slight irregularities in the rock side and a carving probably depicting human-like creatures standing in a row. We could also see diverging fingerlike streaks that might represent an image of a hand, but no consistent picture was possible to identify yet. Bengtsson covered the cliff area with a big piece of paper and taped it in place. Then, on his knees, he started to rub on the cliff with the blueprint paper, which he had rolled around a folded towel (Fig. 5.4).

Then something very special and fascinating happened. As Bengtsson rubbed the paper on the cliff, the contours of two human individuals united by a big, shared hand slowly appeared. In the hand of one of the human-like images, we could discern a sword-like item and the contours of the phallus. The other individual, in a posture of lifted and spread legs, seems to have produced a round object that might be an egg (Fig. 5.4). The whole scene was deeply fascinating and electrifying, I could feel the hairs on my neck stand up. Here we were suddenly watching a greeting emerge from the past – the sacred union!

The rock carvings in Tanum represent an invaluable treasure and have been classified as a world heritage site by UNESCO.

Stones with Hand Signs

Interesting artistic hand images from the past are also found on the so-called hand-sign stones, which are solitary stones with hand motifs (Fig. 5.5). Most of them, about 20 stones, were found in Denmark within a fairly limited area. In Norway there are two such stones with hand signs. In Sweden, three hand-sign stones have been found, all within the parishes of Askum and Tossene in Bohuslän. The first of

Fig. 5.5 The Sanneröd
stone – a hand-sign stone
from the parish of Askum in
Bohuslän (Photo: Lasse
Bengtsson)

the three, the Sanneröd stone, was accidentally found in a farmer's field. A second,
the Svalebacka stone, had been used for more than 20 years as a support for a hen-
house after being unearthed while digging a ditch. The third stone was accidently
found in a stone wall 1993. The hand stones feature an arm with a hand, thumb
extended. On top of the hand are four short transverse lines.

Several attempts have been made to interpret the signs on the hand-sign stones.
Perhaps the four lines represent the main image, and maybe the hand itself has a
general protective effect. Or maybe the lines are there to magnify the importance of
the hand four times. Another possibility may be that the five fingers of the hand plus
the four transverse lines result in nine, the number of months of human pregnancy.
Perhaps the signs therefore symbolise birth.

References

1. Bahn PG, Vertut J. Images of the ice age. New York: Facts on File; 1988.
2. Bahn P, Vertut J. Journey through the ice age. Berkeley: University of California Press; 1997.
3. Snow DR. Sexual dimorphism in upper palaeolithic hand stencils. Antiquity. 2006;80:
   390–404.

4. Clottes J, Courtin J. La grotte Cosquer: peintures et gravures de la grotte engloutie. Paris: Seuil, cop; 1994.

5. Clottes J. Les cavernes de Niaux: art préhistorique en Ariège. Paris: Seuil; 1995.

6. Barriére C. L'art parietal de la grotte de Gargas. Mémoires de l'Institute d'art prehistorique de Tolouse: British Archaeological Report; 1976.

7. Delluc B, Delluc G. Le sang, la souffrance et la mort dans l'art paléolithique. L' Anthropologie. 1989;93:389–406.

8. Leroi-Gourhan A. Les mains de Gargas. Essai pour une étude d'ensemble. Bulletin de la Société préhistorique française Études et travaux. 1967;64:107–22.

9. Lorblanchet M. Peindre sur les parois de grottes. Dossiers de l' Archéologie. 1980;46:33–9.

10. Luquet G. Sur les mutilations digitales. J Psychol Norm Pathol. 1938;35:548–98.

11. Verbrugge A. La main dans la prehistorique. Initiation à l'archeologie et á la prehistorique. 1979;13:25–39.

12. Altuna J, Baldeón A, Mariezkurrena K. L'art des cavernes en Pays basques. Paris: Seuil; 1997.

13. Lewis-Williams D. The mind in the cave: consciousness and the origins of art. London: Thames & Hudson; 2002.

14. Ramos PAS. The cave of Altamira. New York: Harry N. Abrams; 1998.

15. Thiault MH, Roy JB. L'art prehistorique des Pyrénees. Paris: Réunion des Musées Nationaux; 1996.

16. Clottes J. Cave art. New York: Phaidon Press; 2008.

17. Bahn PG. The Cambridge illustrated history of prehistoric art. Cambridge: Cambridge University Press; 1998.

18. Leroi-Gourhan A. Treasures of prehistoric art. New York: H.N. Abrams; 1967.

19. Hodgson B, Blair J. Argentina's new beginning. Nat Geogr. 1986;170:226–55.

Chapter 6
The Intelligent Hand: An Extension of the Brain

Abstract Our hands are essential to us. We take it for granted that they will always be present and ready to intuitively execute our intentions. The hands are intelligent and full of silent knowledge; they remember what they once have learnt. In body language they represent much of our identity. The hand has been called 'the outer brain' and 'the mirror of the soul'. The hands have a special link to the soul. Their 'language', gestures and expressions reflect our mood, whether it is joy, disgust, despair, surprise, thoughtfulness, disappointment, wonder or hope. Hands can appeal, help and welcome. With our hands we can threaten or deny and express empathy and sympathy. We use them to applaud and express approval, wonder or shame, and we can convey quantity and size.

The hand has a rich symbolic value in religion, art and philosophy, but hand gestures may have completely different meanings in different cultures. When the power and function of the hand is not sufficient, it can be extended technologically through a variety of tools such as levers, joysticks and mobile phones. In modern industry, the competence and creativity of hands have largely been replaced by machines and robotic devices, but the hands' accumulated experiences and knowledge are still something genuinely human, giving the hand a key role in society and cultural life.

Our civilisation is primarily based on the knowledge, creativity, abilities and experiences of the human hand. The written word, beautiful handicrafts, a melodious piano piece – they are all a product of creative hands. Handicraft traditions have always constituted a basis for the design and decoration of articles for everyday use. Artistic handicrafts, sculpture and painting are all based on the creative power and silent knowledge of hands.

But we are living in a time when the ability and experiences of the hand are no longer appreciated and utilised as before: In many schools, teaching and training in handicrafts and woodworking are not given sufficient priority. Handwritten letters hardly exist anymore. Handmade objects are becoming more and more uncommon. Today, the hand's abilities are most interesting for handling a computer keyboard, touching the icons on smartphones or handling joysticks to control computer games. In industry, computers have taken over the manufacturing of objects that in earlier times were the product of skilful and sensitive hands.

The late architect Peter Broberg wanted to create an Academy of Hands to take advantage of and safeguard the knowledge and capacities of the hand, with a focus on the handicraft traditions that are still maintained in the area of Österlen in the county of Skåne in southern Sweden. Such an Academy would arrange courses, symposia and seminars addressing subjects associated with the creativity of hands. Several fields and subjects would be represented, such as artistic handicrafts, fine carpentry, blacksmithing, textile art, painting and sculpture, and invited lecturers would discuss and demonstrate the potential of the hand in music and art. Selected parts of courses in anatomy, physiology and neurology would also be included in the activities of the academy.

I think that such a hand academy should have a prominent place in the research community. Without hands, no laboratories would function, no research protocols would have been established and no dissertations would have been written. No computers would be built or used, and no Nobel Prizes would be delivered from the hand of the Swedish king.

Perhaps the hand could be regarded as a symbol of knowledge and understanding. According to the Bible's story of creation, Eve was tempted, despite a very obvious prohibition, to use her hand to pick fruit from the tree of knowledge, to take a bite and then give it to Adam: 'It was a delightful tree since it gave you sense and wisdom and she took from its fruit and ate; and she also gave to her man who was with her, and he ate' (free translation, Genesis 3:6). The punishment was certainly severe – deportation out of Paradise – but the hand had forever conveyed a newborn knowledge to the first humans that was essential for the continuing development of man.

The anatomy of the hand is complex; its function is based on an interaction among muscles, tendons and nerves. But no hands would function without advanced interaction with the brain, so the hand academy would also pursue research into this interdependency. The hand has a cortical representation of its own in the brain and is constantly changing depending on the hand's activities. If the hand is very active and receives high levels of sensory input, its representation in the brain rapidly expands and increases in size. A sensory experience may start with a touch applied to the hand, but the processing and interpretation of the sensory signals happens in the brain. What the hand feels is processed, perceived and understood in the brain.

Movements and gestures performed by the hand have their origins in the brain – the hand is the executive organ, the brain's instrument and executive organ. So where is the beginning and the end of the hand? The hand begins and ends in the brain; the two organs constitute a joint functional unit. The hand has been called an extension of the brain, but perhaps we should also regard the brain as an extension of the hand – giving it access to our minds and souls.

The human hand has always fascinated poets, philosophers and artists. René Descartes (1596–1650) called the hand 'the outer brain', and Immanuel Kant (1724–1804) regarded the hand as 'an extension of the human brain', a link between

Fig. 6.1 Le masque de
Camille Claudel et la main
gauche de Pierre de Wissant.
Number 349, August Rodin,
Musée de Rodin, Paris (Photo
Christian Baraja)

body and soul (Fig. 6.1). Isaac Newton felt that the thumb was 'a proof of God's existence'. For Aristotle the hand was 'the instrument of instruments', a universal tool that can have several functions and can perform a variety of tasks.

It is easy to become impressed by the hands' repertoire: we are dealing with an instrument that can create a violinist's skilful vibrato as well as a pianist's virtuosic finger movements on the keyboard, or mix a pack of cards, throw a dart with high precision, handle a corkscrew, knead a pastry or open a Coca-Cola can. The abilities of the hand range from threading a needle (Fig. 6.2) to a knockout in a boxing ring. The hand can manage heavy lifting (Fig. 6.3) and can pick blackberries, rinse straw-berries and peel potatoes, activities that require sensitivity, fine motor function and coordination. In a social context, the hand represents power as well as love and reconciliation. It can hit and it can caress.

A Mirror of the Soul

The hands and the face are body parts that, in the western world, are freely exposed to the environment and that we like to decorate. In Indian henna painting, the hand is decorated with delightful coloured patterns, and we like to beautify our hands with rings and bracelets. Being a hand surgeon, it is easy to notice how hand cosmetics mean more to some people than to others: for an actor or a sales clerk exposed to numerous customers, the appearance of the hand may be very important, while

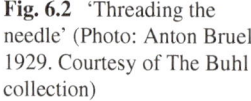

Fig. 6.2 'Threading the needle' (Photo: Anton Bruel, 1929. Courtesy of The Buhl collection)

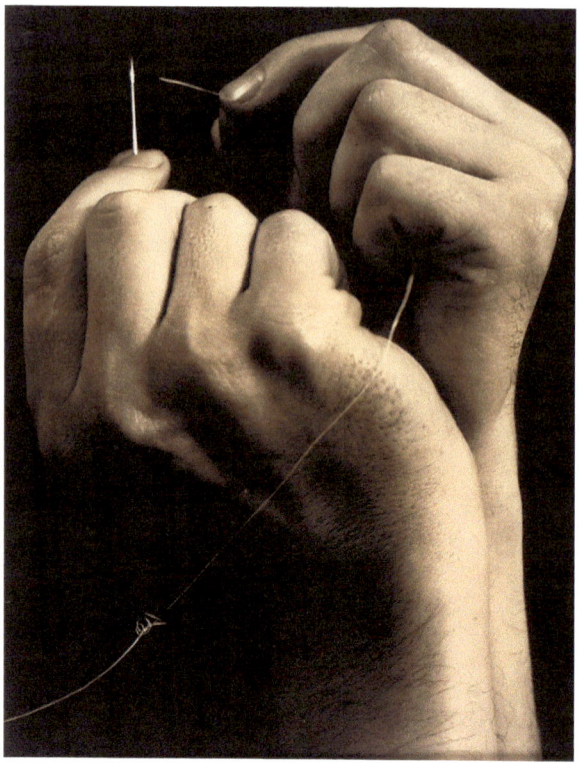

others are more interested in the *function* of the hand than its appearance. In Sweden, the moose-hunting season is more or less magical for many hunters, and several times I have treated patients with hand injuries who refuse to remain hospitalised when moose season is approaching: They do not care what their hand looks like as long as the trigger finger works sufficiently to make shooting possible.

The appearance of a hand reflects much of its owner's current living conditions, and their life and work experiences from past years; big hands with rough skin may indicate a hard life with a concentration on the hands' force and function. It is usually easy to estimate an individual's age by the appearance of their hands. An elderly face can feign youth with the help of cosmetics and face lifts, but the ageing of the hands cannot be hidden.

The hand represents much of our identity, not only by fingerprints but also in body language [1] (Fig. 6.4). Our hands, gestures and movements largely mirror our personality – even if the person's external appearance changes with increasing age or sickness, she can easily be recognised by the way she gesticulates and moves her hands, something that does not change over the years. Identity and personality are mirrored in the handwriting and in every individual's personal signature.

The hands have a special link to the soul; their 'language', gestures and expressions mirror what we feel, wish and want to express (Fig. 6.5). The hand and the

Fig. 6.3 A fisherman from
Österlen in southeastern
Sweden lifts the heavy catch
of the day (Photo: Bo Balldin)

face together mirror our mood – enjoyment, disgust, despair, surprise or fear
(Figs. 6.6 and 6.7), thoughtfulness, disappointment, wonder or hope [2–4]. Hands
can appeal, help and welcome (Fig. 6.8). With our hands we can threaten or deny
and express empathy or sympathy. We use them to applaud and express approval,
wonder or shame. With the hand we can indicate quantity and size.

Hands can bless and heal; the laying on of hands transfers power and force. We
fold our hands in prayer and worship, we clench our fists in anger and we wring our
hands in despair. We use them for defence, but they can also express tenderness and
passion. With a hand over our heart or our fingers on the Bible, we swear the truth.
To point with the whole hand is to signal determination and a strong will.

The hand may symbolise something positive; if something is picked by hand, it
is carefully selected. Handbooks and manuals help us to understand complicated
contexts. An object that is made by hand ensures quality and carefulness. To be in
safe hands makes you feel secure and comfortable. Contact between hands rein-
forces the feeling of solidarity (Fig. 6.9); it makes you feel calm and confident

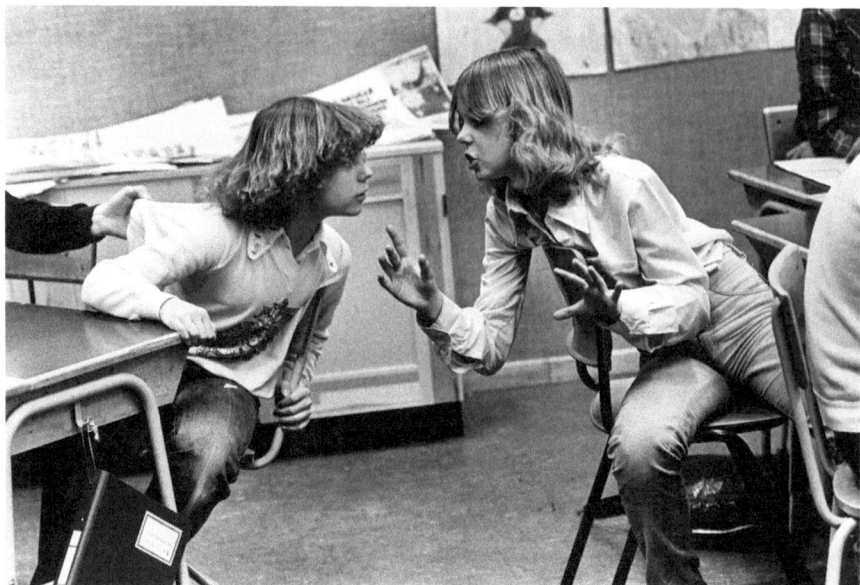

Fig. 6.4 The hand in body language. There is something exciting to tell – not to be discovered by the teacher. Stockholm (Photo: Jan Delden)

Fig. 6.5 The hands are important to enhance our expressions. This man has something important to tell in the Maranatha tent in Stockholm (Photo: Jan Delden)

Fig. 6.6 Sweden takes the lead with 1–0 against Brazil in the FIFA World Cup final in 1958. The Brazilian official can't believe his eyes (Photo: Jan Delden)

Fig. 6.7 Waiting for survivors. The sinking of the *Andrea Doria* 1956. The Italian ocean liner has sunk, and on the quay a crowd of people are waiting for news about survivors of the disaster. A nun gazes into the distance, pressing her hand against her lips and holding her breath – a gesture expressing both hope and fear (Photo: W Eugen Smith. Courtesy of the Buhl Collection)

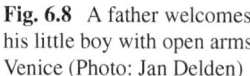

Fig. 6.8 A father welcomes
his little boy with open arms.
Venice (Photo: Jan Delden)

(Fig. 6.10). Holding on to someone gives consolation in a dangerous world – so
does feeling a comforting hand on your head (Fig. 6.11).

The touch of hands is central in close communication between individuals [5, 6].
The hand is there to offer consolation and comfort, friendship, love and sensuality.
In erotic scenes in film and in the literature, the hand may play a major role in inti-
macy and lust (Fig. 6.12).

The Hand and the Art of Medicine

Many in the health-care industry recognise that the hand is intimately linked to the
soul and that injury and illness of the hand may be especially complicated and difficult to treat. The hand is both the instrument and the mirror of the soul, and often
psychological problems may develop with chronic pain conditions of the hand.

Fig. 6.9 When the summer holidays are over, you realise how much you missed your classmates. Stockholm (Photo: Jan Delden)

Biting one's fingernails is a fairly benign sign of stress, but there are examples of psychotic conditions where the patient doesn't feel any limits and literally nibbles and mutilates the fingers. Psychiatric conditions can manifest themselves in other types of self-destructive behaviour, like Secretan's syndrome, where the patient constantly feels forced to induce and maintain chronic wounds on the back of the hand.

For a physician, examining a patient's hand can be a great help in making a correct diagnosis. The character and strength of the pulse may reveal various pathological conditions in the circulatory system including heart disorders. A rapid, thready pulse may indicate a severe loss of blood in severely injured patients. An irregular pulse can indicate cardiac fibrillation with disruptions in the pumping function of the heart.

The physician's own hand may be an important diagnostic instrument for the examiner. It is a universal tool that, when used by a skilful clinician, can make it possible to identify various pathological clinical conditions; this was especially important before the emergence of today's advanced laboratory methods and x-ray techniques. Percussion over the lungs – tapping with the fingertips over the patient's back – may determine if the lungs are filled with air or if there is an infiltration – a dense part devoid of air – which is a sign of infection. Palpating the belly of a patient with severe abdominal pain can be essential for a surgeon's decision as to whether an acute surgical intervention is necessary, as in the case of appendicitis or peritonitis, or if one can wait and follow the development of symptoms. When a

Fig. 6.10 Hand-holding
means security and
consolation – not least in
war-torn areas. Uriel Sinai/
AP/Scanpix

Fig. 6.11 Holding on to
someone gives consolation in
a dangerous world – so does
feeling a comforting hand on
your head. Teheran 1959
(Photo: Jan Delden)

Fig. 6.12 Erotic hands: Tony Leung and Maggie Cheung in the movie *In the mood for love, 2000* (Archive image, Swedish Film Industry)

gynaecologist performs an internal exam, the hand is the instrument that discovers pathological conditions in the uterus or ovaries, and when a surgeon examines the mammary gland, the hand can detect a suspicious tumour. When a urologist palpates an enlarged prostate gland, the finger can indicate whether there is a benign, smooth enlargement or an irregular surface that indicates malignancy.

In the traditional art of medicine, the physician's hand may also have another important role, different from the role as a precision and diagnostic instrument. I am referring to the importance of the physician's hand for conveying confidence, hope and courage. In the book *The Story of San Michele* [7], Axel Munthe, a very popular doctor in Paris and Rome during the end of the nineteenth century, describes his enormous success in the aristocratic community as well as among the poor and penniless. He admits without hesitation that he was not a very skilful physician; he felt that his studies had been too hurried and that his hospital training had been too short. Munthe was quite aware that his ability to instil confidence was the basis for his success and that his hands played a great role in this context: 'I was not a good doctor, my studies had been too rapid and my hospital training too short, but there is not the slightest doubt that I was a successful doctor. What is the secret of success? To inspire confidence. What is confidence? It is a magic gift granted by birthright to one man and denied to another. The doctor who possesses this gift can almost raise the dead.' Munthe was quite aware of the magic power that was conferred on his hands: 'Why do they all obey me, why could I so often make them feel better, even by a mere touch of my hand? Why, even after the power of speech had gone and the terror of death was staring out of their eyes, did they become so peaceful and still when I laid my hand on their forehead? Why did the lunatics in the Asile Saint Anne, foaming with rage and screaming like wild animals, become calm and docile when I loosened their straitjackets and held their hand in mine?' [7].

The Intelligent Hand

The hand is intelligent and teachable; it stores memories and experiences. Once we have learned how to tie our shoelaces or make a knot, the hand can easily do it by itself – its experiences have created tracks and patterns of memories in the brain that are directly linked to the hand. The same holds for the art of tying a necktie or buttoning a shirt.

And the hand helps us to learn. We learn by using the hand – according to Aristotle, you have to *do* in order to understand and remember. When the hand is active, your inner view and perspectives are widened. Knitting a cardigan is easy even if you are watching TV or chatting with somebody – the hand is doing the job by itself. The hand is comfortable and feels at home on the computer keyboard, it remembers where the different letters and figures are located. A piano piece that you have practised many times can almost play itself; it's as though the brain becomes transposed into the fingertips. When we come home late at night and want to unlock the door in the dark, our sensitive fingers can automatically identify the correct key on the key ring – the hand remembers and recognises its specific configuration.

When a hand is active, your thoughts may be far away, beating creative tracks and forming good ideas. When the hand is occupied by its own activities, the imagination may flourish.

The hand can be regarded a sense organ where the sense of touch plays a major role. To feel and perceive shapes and textures, to feel and recognise objects in your pocket or handbag – all this requires an enormously advanced hand sensibility. By the sense of touch we learn to know our nearby surroundings and perceive the shape and texture of the objects that are touched by the hand. In blind people and those with impaired vision, the sense of touch is enhanced – the hand 'sees' what the eye cannot perceive.

In his novel *Revolutions*, Nobel Prize winner in literature 2008 Jean-Marie Le Clézio describes how the young Jean (the author's alter ego) visits his blind aunt Catherine in Paris [8]. Every time they meet there is the same ritual. Clézio describes

Fig. 6.13 Ingmar Bergman directs and seeks the perfect expression during rehearsal of Büchner's *Woyzeck* (Photo: Jan Delden)

how Aunt Catherine reached out her hands until she touched him and how she then slowly moved her hand over his face, how her fingertips followed the contour of his forehead, the brows, the eyes, then the bridge of his nose, his lips and the cleft of his chin. Then she turned the hands around so that her palms were tilted upwards and Jean put his hands in hers without saying a word. He felt that this was a long, quiet and quite dramatic moment [8].

Turning the pages in the morning paper over the morning coffee is an automatic process – we don't have to use our sight or think about how to do it. The hand can even feel the thickness of each page, and it can automatically judge whether we happened to get two pages instead of one between our fingertips.

We take it for granted that our hands will always be there, ready to move intuitively and to be our feelers towards the outer world [9]. Not until we develop problems with our hands do we suddenly realise how important they are and how poorly we would manage without them. Our hands are important in communication with the outer world. Our hands convey messages and signals, and in theatre playing the directors hands are essential (Fig. 6.13). Quintilianus said the hands *can almost speak*. A newborn child points and gesticulates long before it can express itself in words. For people with impaired hearing, sign language is rich and universal. In *Seeing Voices*, Oliver Sacks describes how deaf people, via their visual sense, can experience all details and values of a spoken language [10].

The hand's movements and gestures are naturally linked to speech when we communicate with the surrounding world, even when we are talking on a telephone and no one can see our hand movements. In the western world, shaking hands is a natural component of social interaction. However, we should not forget that the hands also can be dangerous carriers of disease and that contact between hands is an effective way to spread several virus-based illnesses.

The Written Word

The art of writing is one of the big conquests of the hand, allowing us to preserve early written documents, like the Dead Sea scrolls, for thousands of year. The first letter may have been written by a Babylonian 3,000 years ago. Runic inscriptions, done by our ancestors, transmit several-thousand-year-old messages. Preserved handwritten letters can describe historical events and passions that flourished hundreds of years ago. Henry VIII's preserved love letters to Anne Boleyn bore witness to an obsession and endless love that made him break with the Catholic Church and the Pope.

Today, a handwritten letter is a curiosity indicating a very special concern in contrast to today's very brief emails, expecting rapid answers without respite and reflection. There are several examples of handwritten letters being essential for keeping friendship and love alive sometimes over long distances. The love story between the Swedish author Ellen Key and the critic Urban von Feilitzen lived and flourished via thousands of letters during 1880s and the radical following years. The romance lasted a lifetime and was expressed almost only by their exchange of letters – they met on only a few occasions.

The special feeling that is associated with handwriting has been described by several authors, writers and composers. Drawing a pen or pencil over a paper creates a special feeling; the resistance, the friction, the sound of the tip against the paper add a sense quality that in itself is inspiring and could never be experienced by working with a computer keyboard. Even an old typewriter can give such a feeling: a writer I know has told me that he still uses his old typewriter from the 1970s – he needs the feeling of resistance in the keys to find the right inspiration.

Blogs and Facebook have begun to replace letters and diaries, but what sources will remain for historians of the future? Frequent updates of computer systems and programs may make it impossible to read older documents, even after only a couple of decades. Today, I cannot read the documents in my Mac computer that I wrote 20 years ago.

More than 4,000 years ago, the first official post office organisation was created in Egypt. The Egyptian kingdom had grown so complex and large that there was a great need for rapid, secure communication. This could be achieved by the written language – hieroglyphs – and papyrus material. In this way, the couriers of the Pharaoh could bring rolls of papyrus with important messages to all provinces in the kingdom.

The art of writing was known very early among the aristocratic community. For ordinary people, the art of writing letters emerged during the nineteenth century with the introduction of the public school system. The waves of European immigrants to North America in the nineteenth century had many reasons to write to friends and relatives in the old country.

The Symbolic Value of the Hand

The human hand has always played an important symbolic role in philosophy, literature, art and religion. The hand has a central role in several ritual contexts, and the laying on of hands is important in Christian baptism, confirmation and communion. Hands can bless and ban, and we fold them in prayer. In courts, we place our hands on a holy book to swear to tell the truth.

Hand signs and hand gestures possess a strong symbolic meaning, not least in politics. But the meaning of gestures can strongly depend on the context [11–13]. Raised hands, as a Nazi symbol, create great discomfort, but the raised hands of fans at a Bruce Springsteen concert give completely different signals. An example of the power of hand symbolism became obvious at the world at the Olympic Games in Mexico City in 1968 when the two black athletes, Tommie Smith and John Carlos, won gold and bronze medals, respectively. At the distribution of prizes, when the national anthem was played, they both raised their right arms with closed fists in the Black Power sign of the Black Panthers. Smith and Carlos were members of a group of athletes that had joined forces to protest against the treatment of black people in the USA. Among other hand signs with obvious political meaning, the V sign is one of the most well known. Since the Second World War, the V sign has been used by millions of people in many various contexts to symbolise victory or peace. However, it is important that the V sign is performed in the right way, with the palm turned outwards. With the hand turned in the wrong way, the meaning is completely different – the V sign becomes an obscene gesture and a rude insult.

The V sign is primarily linked to Winston Churchill who, during the Second World War, made the sign a symbol for courage and victory. But what is the story behind the V sign? Perhaps Churchill felt that the letter 'V' for victory was obvious enough to instil courage in the people. But there is another version of the origin and the meaning of the sign. A British military historian believes that it has to do with battles between the French and English about 500 years ago when the English archers constituted the true core troops. If an English archer was captured, it is said that the French used to cut off the index and middle fingers on the hand he used to span the bow, thus rendering him permanently useless as a warrior. But at the battles at Agincourt and Crécy in 1415, the English archers defeated the French. When the French war prisoners were taken away under taunts and invectives, it is said that the English archers increased the French pain by triumphantly and scornfully raising their hands with the first two fingers in a V sign.

Hand signs used in everyday life may have completely different meanings in different cultures. If one is not aware of local habits, one risks making a fool of oneself and insulting the local people. The sign for 'OK', forming a ring with the thumb and index finger, can be completely wrong when one communicates with people from other cultures. In France the 'OK' sign can mean 'useless' and 'of no value'. In other cultures, the sign may have an even greater negative or even obscene meaning. It is said that Richard Nixon, as vice president of the USA in the 1960s, made a PR journey to South America in an attempt to thaw the frozen relations between the USA and several South American countries. When he got off the plane on one of the first stops, he made the 'OK' sign to those who came to meet him. The reaction was loud boos, as the 'OK' sign in that country meant 'go to hell'.

In Nigeria, hitchhiking may cause problems, at least if one waves the hand with the thumb extended upwards in the way we are used to. There is a story of an American teenager who tried to hitchhike while on holiday in Nigeria. A car passed by and came to a sudden stop. The hitchhiker was beaten up. In Nigeria a thumb raised upwards is a rude insult. In contrast, in other cultures, 'thumbs up' is usually a sign of encouragement and a positive agreement. 'Thumbs up' was once the Roman emperor Julius Caesar's sign that a defeated gladiator was pardoned.

Extending the Hand Through Technology

The hand is a prerequisite for technical development. When the hand's force, gripping function and creativity are not sufficient, we enhance its capacity by developing technical products, machines and robotic devices. When the hand cannot lift a heavy concrete block, we extend it with a lever or we train it to manage a joystick, controlling the movements of a crane, and when a shovel is not enough, we extend the hand with the controls of an excavator machine. By using such technical extensions and amplifiers, the biotechnical human can go beyond nature's limits. When man once learned how to plant a seed or a plant by hand, the ground's capacity to produce food was amplified several times.

The hand plays a key role when we use computers to communicate with the outer world, manage our finances via internet banking or surf the internet. With a mobile phone in our hands, we can instantly reach any place on the earth. In an interview in a Swedish newspaper, an African woman expressed fittingly how the hand can be extended by this technology: 'I always have my mobile phone in my hand; it is almost like an extra body part'. The woman lived in Tanzania in an area with no ordinary telephone system. Time is reduced to a minimum when an email to the other side of earth can be answered within a few seconds; the exchange of letters as in times past, requiring days and weeks, is almost completely gone and has been replaced by rapid communication on the internet.

In our daily lives it is important to remember the correct PIN code when using a cash machine, opening a locked door or filling your tank with petrol. However, this process requires coordination of the interaction between hand and brain – one has to remember the combination of figures and transpose this memory into pushing the right buttons on the display. Sometimes the hand remembers the positions of the buttons on the display rather than the actual figures – which can cause big problems if the figures are in a different order, for instance when you use a cash machine in a foreign country. The memory tracks of the eye and the hand may conflict.

The hand can be extended by technology and may sometimes be replaced by various mechanical innovations, but handiness and dexterity are genuinely human and based on knowledge, creative power and experiences of the hand. This gives the hand a key role in culture and in a flourishing, diverse society. The knowledge and abilities of the hand are still necessary when it comes to skills such as handling a scythe when harvesting and maintaining a flower meadow: a correct, sharp cut through the flower stems creates conditions for regrowth of the special flora, while a mechanical trimmer demolishes the vegetation so that regrowth is inhibited. The hands cannot always be replaced by machines: if there were no hands, there would be no carpenters, sculptors, engravers, ceramists, musicians, illusionists, jugglers or farriers.

References

1. Wing AW, Haggard P, Flanagan JR. Hand and brain: the neurophysiology and psychology of hand movements. San Diego: Academic; 1996.
2. Erwitt E. Elliot Erwitt's handbook. New York: The Quantuck Lane Express; 2003.
3. The Buhl Collection. Speaking with hands, photographs from the Buhl Collection. New York: The Solomon Guggenheim Foundation; 2004.
4. Verdan C. La main cet Univers. Edition du Verseau, Roth and Sauter ed. Denges: Fondation du Museé de la Main CH-1026 Denges, et Fondation Claude Verdan, CH-1005 Lausanne; 1994.
5. Paterson M. The sense of touch: haptics, affects, and technologies. Oxford: Berg publishers; 2007.
6. Josipovici G. Touch. New Haven: Yale University Press; 1996.
7. Munthe A. The story of San Michele. London: John Murray; 1929.
8. Le Clézio JMG. Révolutions. Paris: Gallimard; 2003.
9. Wilson FR. The hand: how its use shapes the brain, language, and human culture. 1st ed. New York: Pantheon Books; 1998.
10. Sacks O. Seeing voices: a journey into the world of the deaf. London: Picador; 2011.
11. Axtell RE. Gestures: the do's and taboos of body language around the world. New York: Wiley, cop; 1998.
12. Morris D. Gestures. Their origins and distribution. A new look at the human animal. London: Granada; 1982.
13. Stam G, Ishino M. Integrating gestures: the interdisciplinary nature of gesture. Amsterdam: John Benjamins Pub. Co.; 2011.

Chapter 7
Touch

Abstract The hand is a sensory organ as important as the eye and ear. The sense of touch is essential for exploring the surrounding environment, and the tactile sensibility of the hand easily identifies textures and shapes. Without tactile sensibility, the hand loses most of its function. Hand sensibility can replace vision in complete darkness. Vision shows us the shape of objects, while the sense of touch confirms the impression and gives a more complete picture of its consistency, hardness, material properties and texture, even on the backside and inside – *seeing is believing, but touching is understanding*. While the hand's tactile sensibility is essential in most everyday tasks, *protective sensibility* is essential to avoid harmful burns and wounds. Sensory feedback is essential for regulating grip force. A body part that loses sensation and sensory feedback triggers a lost sense of 'body ownership' – it is no longer perceived as part of the body. Congenital insensitivity to pain is a very uncommon illness found among some families, especially in northern Sweden.

Calling the hand a sense organ may sound strange since it is usually the eye or the ear that are regarded as obvious sense organs. Touch is but one of our five senses, the others being sight, hearing, smell and taste. Problems with vision and hearing are somewhat common, but problems with hand sensibility are not equally well known. We take it for granted that the tactile sensibility in our hands will always be good and that our hands will function normally. We expect our hands to be able to estimate the temperature of the water in the shower by just feeling the stream. With our sense of touch, we can decide whether a shirt is made of cotton or synthetic material, and we can estimate the character and structure of the textile by feeling its texture. With our fingers, we easily detect tiny scratches as well as small crumbs from bread and grains of salt on the breakfast table, and we have no problem handling a fork and knife or a drinking glass or picking up a lump of sugar even if our sight never leaves the text in the morning paper.

But if the fingers lose their sensibility, they also lose their fine motor properties. Everyone who has lost feeling in their fingers on a cold winter day knows that it can

G. Lundborg, *The Hand and the Brain*,
DOI 10.1007/978-1-4471-5334-4_7, © Springer-Verlag London 2014

be totally impossible to grip a key and lock the door. A hand without sensibility is usually also a hand without function.

The sense of touch and the fine sensibility of the hand makes it a sense organ, equally important as the eye and ear [1–3]. We are dependent on hand sensation in most activities of daily life in our homes, at work and at leisure. And of course when we spend time together with our family and closest friends, how would we be able to caress the hand and cheek of our loved ones with a hand without sensation?

Normally hand sensibility and vision cooperate and interact when we experience the world around us, but in certain situations, hand sensibility plays a crucial role. When there is a sudden power failure at home on a dark winter night, our hands immediately take on the role of feelers towards the surroundings. Without vision we grope along walls and well-known features of our home until we ultimately find a drawer, explore its contents and identify a candle and a match to light it – not an easy task for two hands in total darkness.

The hand is a master when it comes to finding and identifying objects in total darkness. One dark evening in September, I finally reached the mountain lodge I was hiking to on a rainy day in the mountains of northern Sweden and could finally take off my backpack. There was no electric light in the cabin and no paraffin lamp. I had to quickly pick up a change of clothes from the deep layers of the backpack – socks, underwear, a shirt and a sweater. My hand searched in complete darkness and had no difficulty identifying and picking up all these objects: In the darkness my hand could 'see' what was hidden. My hand could easily differentiate between stockings, socks and shirts, and it was irritated by non-visible small scraps from a damaged package of biscuits in the bottom of the sack.

The role and importance of the sense of touch and the tactile sensibility of the hand has been described in a very poetic, sensitive and accurate way by the Swedish poet, novelist and 1974 literature Nobel Prize Laureate Harry Martinson in his poem 'Human Hands' from 1971, published in the poem collection *Dikter om ljus och mörker* (poems about light and darkness):

The experience of hands is tactile.
Their life among things is manifold,
filled with silent contents.
They do not hear, but sense vibrations.
They do not see, but know how it is in dark cellars,
when velvet is to be valued they are there,
and silently they test the grindstone and the scythe's edge.
No need to let the edge bite down.
With a light touch they feel the steel's sharpness.
How have they found time to collect all their fine experiences
of wool and gravel, of down and steel,
of smooth surfaces and prickly thistle-heads,
of supple talcum and of every kind of flour.
Their range is immense
from shiny silk to coarse sacks,
from rough files and graters
to the smooth nails of the new-born
and the touch-shine on everlasting flowers.
They live in the land of feeling where touch is everything

and where the mystery of touch is the bridge between nerve and soul.
But they find their limit in the scales of the butterfly's wing.
Translated from the Swedish by Judith Moffett and Lars-Håkan Svensson

In poetic colours, Martinson describes the sense of touch and the inborn potential of the fingertips' fine-tuned sensitivity. The sensory experience is based on the accumulated knowledge and previous experiences of the hand; recognising shapes, an item or the character of a texture requires memories of earlier sensory experiences – *the experience of hands is tactile*. The hands' abilities are based on ingrained programming to perceive and understand forms and textures, a learnt capacity to estimate and perceive shapes and forms, which dates back to the newborn child's first exploration of its nearest environment. The newborn child uses all of its senses and the most sensitive parts of its body, its lips and hands, to explore and touch the shapes and textures of items. Sensory and visual inputs interact, and experiences and memories are collected by simultaneously touching, observing, tasting and smelling objects in the vicinity, forming the basis of a lifelong ability to perceive and understand forms and shapes.

Vision and touch interact effectively when exploring the environment; hearing and touch have much in common – *They do not hear, but sense vibrations*. Hearing and tactile sensibility are based on vibrations transmitted via a tympanic membrane (hearing) or via irregularities of the fingerprint ridges inducing vibrations when the fingertips are moved across an irregular surface. Martinson is well aware that tactile sensibility is based on vibrations, even if it is a matter of silent vibrations.

Martinson also describes how the sense of touch guides us when we proceed in darkness – *They do not see, but know how it is in dark cellars*. With the fingers we feel irregularities of the rough walls in dark cellars, perhaps defects in the paint and perhaps also the moisture of the mortar. The hands even feel the stickiness and the faint elasticity in the fine threads of a network of cobwebs. The hand can 'see' around corners, and both hands simultaneously 'look' in various directions when we find our way in the darkness.

Martinson describes the immense range and the hands' ability to recognise and feel the character of textures: *Their range is immense*. Through the sense of touch, we experience the character of wool and gravel, down and steel, even supple talcum and every kind of flour. The sense of touch makes it possible to differentiate various surfaces and textures, everything from coarse sacks and rough files to shiny silk and the smooth nails of the newborn. The sense of touch makes it possible to judge the quality of cloth and textile: *When velvet is to be valued they are there*. Martinson puts a limit on the sensitivity of the hand: *But they find their limit in the scales of the butterfly's wing*.

The ability to feel the structure and character of a texture can be especially important in certain occupations. Vera, aged 65, has worked at a hand bookbinding company for most of her life. I get in touch with her because I need to recondition a copy of Charles Bell's *The Hand, its Mechanism and Vital Endowments*, a seminal work from 1834 about the anatomy of the upper extremities of various animal species, and the role of the hand as a sense organ in man. Unfortunately, the binding had come apart and several pages were loose. Stroking her fingertips over the paper and the bindings, Vera tells me how she can tell simply by touch the quality and thickness of various sorts of paper and how calfskin, goatskin and whale skin bindings differ in terms of roughness, friction, suppleness and elasticity. Her hands' ability to experience all this had disappeared some years earlier when she developed carpal tunnel syndrome (a nerve entrapment at wrist level), but the tactile sensibility of her hand recovered following a simple surgical procedure decompressing the nerve.

Martinson also touches on another important sensory function in the hand: sensation as a protective mechanism to avoid hand injuries. Prickly thistle-heads can injure the hand, but the pain sense warns and puts a limit on what the hand can tolerate. The sharpness of the scythe's edge can be anticipated without the hands being hurt by the sharpness. The sense of touch and pain interact to protect the fingers: *No need to let the edge down. With a light touch they feel the steel's sharpness.*

Martinson is aware that the hands represent a link between body and soul, especially regarding feelings and tactile discrimination. He writes of the hands: *They live in the land of feeling where touch is everything and where the mystery of touch is the bridge between nerve and soul.* He knows that it is not always easy to differentiate between tactile sensitivity and feelings. To *be touched* can be a matter of touching the heart rather than touching the skin, and a touching experience leaves an imprint on the soul rather than on the hand. To lack the *right feeling* indicates a lack of insight and understanding. In intimate social interactions, touch is a basis for tenderness and solidarity. A touch of hands represents friendship, love, sensuality, affection and sexuality.

In Anglo-Saxon literature, the word touch is used in many contexts in the border zone between tactile sensibility and feeling [2–4]. *Let's stay in touch* has a deep meaning in terms of continuing contact, interaction and solidarity. To find an event *touching* is an experience that touches the heart.

Tactile sensibility and feelings interact in the contact between hands. Contact between hands represents security, and a friendly or tender touch of the hand creates a feeling of comfort. The skin possesses a special system of thin nerve fibres that react to slow stroking and caresses, transmitting a feeling of comfort, pleasure and delight. Such nerve fibres do not project to the usual areas of sensory input in the brain cortex, but rather to the limbic system, an area of the brain that is associated with feelings of enjoyment and comfort [5]. It is well known that skin-to-skin contact, not least caresses and contact between hands, can induce such effects. This may explain the beneficial effects of 'tactile massage' and the calming effect that hand contact and stroking over the back of a hand may have in states of unease and anxiety, especially in elderly people.

The feeling of pleasure associated with skin stimulation and the touch of hands may be associated with a release of the hormone *oxytocin*. Oxytocin was first described in breast-feeding mothers, but it is present in all of us and plays an important role in the feeling of well-being associated with hand contact between individuals as well as stroking the fur of animals. The release of oxytocin is stimulated not only by body contact but also by other pleasant experiences like sun, warmth, pleasant music and enjoyable moments with friends. Oxytocin stimulates the limbic system – the brain's 'reward system' – and thus may stimulate feelings of enjoyment and pleasure

Fig. 7.1 Seeing fingertips
(From Erik Moberg *Akut
handkirurgi*; Lund 1969)

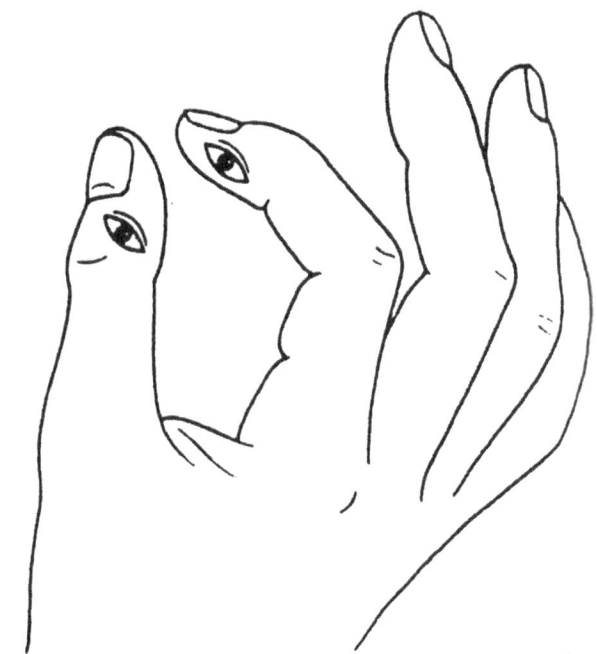

The Seeing Hand

The refined sensibility and delicate motor functions of the human hand make it an exten-sion of the brain towards the outer world; Immanuel Kant even called the hand 'the outer brain'. The tactile sensibility gives 'eyes' to the hand, creating the ability to produce an inner true picture of the environment through the act of touch [3, 6, 7] (Fig. 7.1).

David Katz (1884–1953), an authority in the field of perception psychology, had a specific interest in the interaction between hand sensibility and sight [2]. Katz describes how the eye can see the outer characteristics of objects, while the hand can see the inner characteristics. Touch tells us the true character of an object: Sight reaches the surface and gives us an impression of the shape, while touch confirms the perception and creates a full picture of the object's consistency, hardness, mate-rial properties and dexterity, even on the backside and inside (Figs. 7.2, 7.3, 7.4, and 7.5). The sense of touch helps us to understand and explore the vicinity: 'seeing is believing, but touching is understanding' (Fig. 7.6).

The hands' *tactile* (from the Latin *tactus*, touch) properties refer to the ability to perceive touch (passive tactile perception). But when we use our hands to examine an object, we do so through exploratory movements of the fingers. We feel around corners and edges, perhaps we roll and turn the object between our fingers to get an idea of its shape. This is called *haptics* (after Greek *aptesthe* – to touch), active tactile perception, when sensitivity and motor functions work together actively exploring objects.

The difference between passive and active touch has been discussed by Gibson, who regards the movements of the hand when examining an object as a correlate to

Fig. 7.2 The tactile sensibility of fingertips makes it possible to perceive the shape and texture of small objects

Fig. 7.3 With fingertips we can perceive the fine structure of a dandelion seed head

Fig. 7.4 With the fingertips we can feel the finest detail of a coral's superficial pattern (From Lundborg [13])

the movements of the eyes scanning the environment [4]. He describes 'the act of touch' as the hands' exploring movements to get a perception of an object. The hands' ability to feel pressure and vibrations through such activities is based on a complex interaction among several various types of mechanoreceptors in the skin of the fingers, something that is further discussed in the next chapter.

Fig. 7.5 With the fingertips we can perceive the brittleness of the fibre structure of a coral (From Lundborg [13])

Fig. 7.6 'The act of touch' is important for exploring and understanding the world around us. Hans-Peter Feldmann's photo depicts a surrealistic situation in which a little girl uses her index finger to try to understand the nature of the figure before her. Is it a real girl or just a fantasy figure that somehow casts a shadow? Feldmann leaves it to the observer's imagination to solve the mystery. The artist has intentionally cut out the other girl's figure from the photo

A special system of nerve fibres responds to cold and heat; the temperature of the objects is important, making it possible to differentiate between them as with metal and wood.

The most basic of the hand's sensory functions is the ability to feel pain; the *protective sensibility* of the hand makes it possible to avoid harmful objects that might prick, cut or burn it. With regard to the *function* of the hand, the ability to feel pressure, vibration and touch is of course a prerequisite for more advanced sensory functions. *Localising* the touch of the hand and differentiating several synchronous and closely situated touch points requires greater processing in the brain, especially if the touch points are very close to each other. This type of sensory quality is usually named *discriminative sensibility*, defined as the ability to distinguish between two very nearby pressure points applied to the skin. Normally, the limit for such discrimination between two pressure points – the two-point discrimination capacity – in the fingertips is 2–3 mm. However, on the back of the hand or on the forearm, it is quite different; the two-point discrimination capacity is usually much poorer, a distance of several centimetres.

An even more advanced sensory function is the ability to identify form, surface, structure and texture, which required a high density of sensory receptors in the skin as well as an advanced ability of the brain to process the nerve signals from various parts of the hand [8]. Working in tandem, the hand and brain make it possible to perceive and understand such properties as shape, size and temperature as well as consistency, hardness, wetness, dryness, elasticity, brittleness, stickiness, oiliness and toughness. Sensory perception can be a complex product of sensory stimuli: the perception of wetness may be a combination of temperature and pressure; consider how we check the moisture in a flower pot with a finger. Oiliness can be a combination of resistance and smoothness; you need to move your fingers to feel the oiliness. The hand can perceive the character of pouring water and even the speed of the flow, or the feeling of a faint breath of air against the skin.

The hand's ability for three-dimensional sensitivity, *stereognosis*, is a property that differentiates man from most animals and is linked to stereoscopic vision and a well-developed brain. In the 1960s, Erik Moberg, one of the true pioneers of hand surgery, popularised the term *tactile gnosis* as a term for the fingertips' fine, high-resolution sensitivity that 'gives the fingers eyes' [9–11].

Sensory Feedback

Sensory feedback from the hand is important for controlling the hand's grip force: If we do not know how powerful our grip is around an object, we cannot control the force of our grip [12–15]. Sensory feedback is based on a complex interaction between sensory functions and muscle functions. This feedback is necessary for the hand to also act as a fine motor instrument with a precision grip; otherwise, how would we be able to grasp a bird's egg without crushing it or dropping it? And how would we be able to pick a ripe raspberry without mashing it with our fingertips or pull a grape from the bunch without crushing it? (Fig 7.7). The sensory feedback from the fingertip is crucial when we press the camera's trigger to take a photo at the right moment. With the finger on a rifle's trigger, we depend on sensory feedback to shoot at the right moment.

Striking a match to light a cigarette is another example of a complicated hand movement requiring good sensory feedback and appropriate interaction between the hand's sensitivity and motor function: The match has to be struck with appropriate force against the matchbox's striking surface without breaking the match, and the hand needs to move the burning match to the cigarette at just the right speed so the air resistance doesn't extinguish the flame (Fig. 7.8).

Fig. 7.7 Removing one grape from the bunch without crushing it between your fingers requires perfect sensory feedback

Fig. 7.8 Striking a match to light a cigarette requires good interaction between the sensory and motor functions of the hand. The match needs to be struck against the matchbox without being broken, and once it flares, the burning match must not be moved so fast that the flame dies because of the air resistance. In the movie *Breakfast at Tiffany's*, George Peppard tries to light Audrey Hepburn's cigarette with a match, but Martin Balsam is faster – he use a cigarette lighter which is simpler and quicker to handle (Archive image, Swedish Film Institute)

Proprioception, or our awareness of the location and position of our limbs, also plays an important role in such a movement. Without looking, the proprioceptive sense will tell us the position and posture of the hand. The proprioceptive sense is based on sensory receptors in muscle, tendon and joint capsules. The sensory receptors in the skin also play an important role. When the hand is closed, the skin on the back of the hand is stretched, stimulating the sensory receptors that are activated by tension. In this way sensory feedback from the hand tells us how far the hand is closed [16, 17].

Sensory feedback from the hand and arm interacts with experiences and learned motor patterns for balancing the gripping forces of the hands and arms in various situations, but the system can quickly be recalibrated. Lifting an unopened litre carton of juice from the breakfast table requires a certain force in the hand and arm. By experience, you know intuitively how much lifting force is needed. However, if, during the night, someone consumed half the contents of the carton without your knowledge, your initial pre-calibrated lifting force may be far too strong, and you'll lift the carton much faster than expected. But thanks to an immediate adaptation of the hand-arm system, the lifting force is almost immediately recalibrated and reduced to an appropriate level.

Sensibility, sight and proprioception interact in several common activities in our everyday life. Frying an egg can be seen as a simple task, but from the hand's viewpoint, it is quite complicated: The hand picks up an egg without either dropping it or squeezing it to the breaking point. Then it taps the egg against the edge of the frying pan with sufficient force to crack the shell so that, in next movement, the hands can open both halves of the shell and pour the contents into the frying pan.

Here is a situation where no vision at all is needed: When you suddenly feel a mosquito bite on your leg, your hand can immediately localise and identify the affected area, feel the slight swelling in the skin and ease the itch by scratching and massaging the area.

Sensation and Balance

Keeping one's balance is not always easy, especially while walking on uneven ground. Walking down a staircase without support can be difficult, particularly when the sharpness of the senses has faded with increased age. The balance function is a complicated process that depends on cooperation and interaction between the balance organ in the inner ear and several senses like vision, hearing and sensibility in the soles of the feet. Hand sensitivity can also play an important role in balance, for instance when walking down a staircase. Keeping a hand on the handrail can be very helpful for maintaining balance as we proceed down the staircase, and often just a very light touch with one finger is enough. It has been demonstrated that the sensory information transmitted via the contact between the handrail and the pulp of a little finger can be enough to give the body correct balance. The slightest movements and indentations on the skin of the finger are enough to inform the brain about changes in body position, allowing you to change your posture to maintain balance [18, 19].

Sensation and Body Ownership

In his book *A Leg to Stand On*, the British neurologist Oliver Sacks describes a dramatic event when he was walking in the mountains on top of the Geirangerfjord in Norway [20]. Suddenly the author met a sign reading 'beware of the bull'. Beyond a little hill, when he spotted the horrifying bull, it created a panic reaction and Sacks fell down the hillside. He injured his left leg severely. Being a true Brit, he had brought an umbrella, which he used to splint the leg. After some time, Sacks was transported to a local hospital in the nearby city of Odda where it was discovered that he had lacerated a thigh muscle and had a very serious nerve injury in the leg. He was brought by air to London where he had to undergo surgery.

As a result of the injury, Sacks had completely lost all sensation in his left leg, something he described as enormously frightening and shocking. Due to the loss of sensibility, his leg no longer felt like part of his body. After waking up from the general anaesthesia, when he lifted the quilt to look at the newly operated leg, he made a horrifying discovery: The leg was no longer there. It had slipped off of the bed and was hanging down at the bed side. Sacks had no control over the leg – he could neither move it nor feel it. The contact was broken and he did not perceive the leg as a part of his body anymore; it was transformed into a foreign object. Sacks felt as if his leg had been amputated and had lost its connection with his body and brain. Sacks had not only lost all sensory functions *in* the leg but also all feelings *for* his leg; he was suddenly denying a part of his own body – 'I had lost my leg'.

Most of us have probably had a similar experience when an arm has been under pressure while we slept and we awake with a completely numb hand with no sensation. It is a very unpleasant and frightening experience that suddenly transforms the hand into a foreign object – one that is anatomically connected to the body but no longer feels as if it belongs to the body. The numb hand can no longer grip an object and cannot perform any movements, and we can pinch it without feeling anything at all – the hand is completely numb and paralysed. When sensitivity in a body part disappears, the feeling of ownership of it also disappears, even if our eyes tell us that the hand is there.

Interestingly, with training, hand sensitivity can be transferred into a foreign object that can then be perceived as a part of us – an extension of the hand. When we cut up meat on our plate, the knife and fork can become extensions of our hands; we feel when the knife passes through the crisp surface, and we experience the softness or toughness of the meat because pressure and vibrations in the knife or fork are transmitted to the sensitive skin in our hands. The result is an illusion that the knife and fork are extensions of our hands. The same phenomenon is true when we enjoy Chinese food using chopsticks – the sticks become extensions of our hands.

An experienced carpenter no longer considers his drill a foreign object but rather an extension of his hand – a result of the vibrations and movements of the tool being transmitted to the sensory receptors of the hand. In the same way, the white cane of a blind person becomes an extension of her hand, 'feeling' surfaces and irregularities on the ground. Even a car seat, combined with visual impressions and sounds from

Fig. 7.9 The protective sensibility of the hand is important to avoid injury – the sharp thorn of a cactus can easily damage the skin

the motor and the tyres, may make an experienced driver feel that the car is a part of himself and that his intuitive movements of the steering wheel and pedals are as natural as his movements of arms and legs during an evening walk. Perhaps the illusion of the car's connection to the body is a prerequisite for manoeuvring it intuitively and safely. The hand can be extended by all sorts of everyday tools, for instance, when we write with a pen, whip cream or turn pancakes. None of these activities are performed with the hand directly, there is a tool in between. But the hand receives sensory feedback from the tool and is therefore interpreted as a part of the body.

Several studies have shown that the use of a tool that is held in the hand may also influence the representation of the hand and arm in the brain: the arm representation in the sensory brain cortex is changed to also include the tool [21–23]. The person using the tool feels as if her arm is extended, and her brain views the tool as part of her extended arm – the tool has its own representation in the brain. When we brush our teeth, the toothbrush becomes a part of us and an extension of our hand, an illusion that is further reinforced by the toothbrush touching very sensitive parts of the body – the mucous membranes of the mouth. A musical instrument can become a part of the body: a violinist can feel that the violin and the bow become extensions of his arm. Such an illusion is probably a prerequisite for an optimal performance when the musician has mastered the technique.

Protective Sensibility

The most basic and fundamental sensory function in a hand is protective sensibility, the ability to feel pain induced by exposure to stimuli that are harmful to the hand, so-called nociceptive stimuli (Fig. 7.9). Without this protective sensibility, the hands can easily be injured. Obvious examples are patients with nerve injuries who lack sensitivity in the hand and fingers. If such a patient smokes a cigarette, the glowing embers can easily reach the fingertips without the patient's notice until he smells the burnt skin. Thus, the ability to feel pain has the essential function of protecting the hands from injury.

Pain: The Gift Nobody Wants

The legendary surgeon Paul Brand was trained in England but worked the major part of his life in India, where he treated people suffering from leprosy. In 1993 Brand and Philip Yancey published the classic work *Pain: the Gift Nobody Wants* [24]. In the book Paul Brand describes how the ability to feel pain protects us from injuries and wounds on our hands, feet and face. He posits that pain can be a loyal friend if we learn to listen to it and to react in the right way.

Leprosy is a special type of nerve lesion where increasing swellings in the nerve trunks destroy nerve fibres so that the ability to feel pain – protective sensibility – disappears. The result is that the people suffering from leprosy easily burn their hands, and wounds easily occur on the feet and other exposed body parts like the nose and ears. The situation can result in spontaneous amputations of fingers and toes. Malformed and very poorly functioning hands are common with this condition.

Paul Brand understood that the spontaneous amputations of peripheral body parts often seen in tropical countries were associated with the habit of sleeping on the floor and that toes and fingers lacking sensitivity were attractive to smaller rodents. This explanation took the sting out of the myths and the stigmatisations associated with the disease – that amputations were God's punishment.

Leprosy is often mentioned in the Bible and is frequently associated with horror. A leper was regarded as unclean, his clothes would be burnt and he would be isolated outside the city. The disease has always been associated with horror and shame. In Greece, lepers were exiled to the isolated rock island Spinalonga outside Crete as recently as the 1960s.

Leprosy is one of the oldest known diseases. At the time of the birth of Christ, leprosy was common in the Mediterranean area, and it has existed in Europe since medieval times. In thirteenth-century Europe, there were more than 20,000 nursing homes for lepers, and every large city had a leprosy hospital situated outside its border.

In the Scandinavian countries, leprosy became a common disease [25]. The last leprotic woman in Sweden, Kristina Asplund, was born in 1887 and died in 1976. She spent a great part of her life in a leprosy hospital in Järvsö, in the northern part of Sweden, which closed in 1943. In a TV interview, Kristina Asplund had a chance to describe the problems associated with loss of hand sensibility: 'I have no outer sensibility, I cannot feel whether it is warm or cold'; 'I cannot use my hands, I cannot help myself and cannot dress myself'. Kristina Asplund, who became totally blind, also described how the illness started: 'Late in the summer of 1918 I was sitting in a chair knitting stockings. I noticed that my fingers were not as quick and nimble as before. My hands felt strangely lifeless. Yes, they were almost without sensibility'. Regardless of Kristina's hands being covered by burns and wounds and her hands being malformed and numb, she served as a maid. At that time her hands were severely deformed by wounds and burns because she didn't feel anything when she touched something hot in the oven. 'Suddenly there was a moment of pain, there was a burnt smell and then blisters occurred producing fluid and pus'. A doctor's report from 1928 reads: 'Left hand, forearm and fingers are severely deformed. Much fluid is pouring from an open wound in left thumb. As a protection she has only a strip of fabric' [25].

Kristina Asplund's spectacular story tells us in a lively way about the conditions of people suffering from leprosy, and how loss of protective sensibility may have severe consequences for the well-being and function of the hands. But Kristina Asplund was not alone with her problems in Sweden. There are descriptions of how the lepers were accepted and incorporated in society in a way that was not common in other parts of the world. Before the leprosy hospitals were built in Sweden at the end of the twelfth century, lepers seemed to be allowed to move freely and participate in farmwork, working among other people.

Vittangi Disease: Congenital Insensitivity to Pain

In the movie *The Girl Who Kicked the Hornet's Nest* [26] based on the last part of Stieg Larsson's Millennium trilogy, there is a noteworthy scene where Lisbeth Salander nails the blond giant Ronald Niedermann to the floor without him feeling the slightest physical pain. Several previous sequences had already shown Niedermann subjected to very forceful violence without feeling any pain at all. The explanation is probably that Niedermann was not able to feel pain – he might have suffered from a very uncommon 'illness' – congenital insensitivity to pain.

Perhaps Stieg Larsson was aware of, and had been inspired by, Vittangi disease, a congenital illness that results in insensitivity to pain [27]. For several years, Gällivare Hospital in northern Sweden had treated a large number of patients with severe fractures in their feet and lower legs, who felt no pain. The orthopaedic surgeon Jan Minde became interested in the phenomenon and began to investigate whether there was any familial relationship among these fracture patients. He found that they all descended from the same family: Hindrich Mickelsson Kyrö and his wife who, in 1674, arrived in the area where the Vittangi and Torne rivers meet. They founded the community of Vittangi, situated about 70 km south of Kiruna in northern Sweden. Today, Vittangi has about 1,000 inhabitants.

Jan Minde's studies resulted in a thesis at Umeå University in 2006: *Norrbottnian congenital insensitivity to pain* [28]. He found that people suffering from the Vittangi disease had an impaired ability to protect themselves from injury since they could not feel pain. Their bones were easily fractured because they overloaded their joints, and they frequently suffered burns, bruises and cuts [29]. These people sometimes displayed self-destructive behaviour: One girl found it very enjoyable to jump from the wall bars in the gym onto the floor right on her kneecaps because she liked the funny cracking sound when she landed on the floor.

Minde's team at Umeå University showed that congenital insensitivity to pain was linked to a specific gene in chromosome 1 where there was a mutation of one of the growth factors, nerve growth factor beta (NGF beta) [30, 31]. This growth factor is necessary for the development of pain-transmitting nerve cells in the dorsal root ganglia and their nerve fibres. Those with the most severe forms of congenital insensitivity to pain had inherited the predisposition from both parents, while those who were less severely affected had received the predisposition from only one parent. Tissue samples showed that the nerve trunks of those who were insensitive to pain contained fewer pain fibres [30].

There are several types of congenital insensitivity to pain [30]. I once operated on the hand of a woman from northern Sweden with congenital insensitivity to pain. No anaesthesia was needed. The woman told me that there was only one skin area where she could feel pain and temperature – the outside of her right upper arm. To adjust the shower temperature, she had to start out by putting the outside of her right upper arm into the water stream – something she had become used to and that didn't cause her too much trouble.

References

1. Josipovici G. Touch. New Haven: Yale University Press; 1996.
2. Katz D. The world of touch. Hillsdale: L.Erlbaum; 1989.
3. Paterson M. The sense of touch: haptics, affects, and technologies. Oxford: Berg publishers; 2007.
4. Gibson JJ. Observations on active touch. Psychol Rev. 1962;69:477–91.
5. Olausson H, Lamarre Y, Backlund H, Morin C, Wallin BG, Starck G, et al. Unmyelinated tactile afferents signal touch and project to insular cortex. Nat Neurosci. 2002;5(9):900–4.
6. Heller MA, Schiff W. The psychology of touch. Hillsdale: Lawrence Erlbaum Associates; 1991.
7. Diderot D. Letter of the blind, for the benefit of those who see, trans M J Moran. In: Morgan J, editor. Molyneux's question: vision, touch and the philosophy of perception. Cambridge: Cambridge University Press; 1977.
8. Kandel ER, Schwartz JH, Jessell TM, Siegelbaum SA, Hudspeth AJ. Principles of neural science. 5th ed. New York: McGraw-Hill; 2013.
9. Moberg E. Objective methods for determining the functional value of sensibility in the hand. J Bone Joint Surg Br. 1958;40-B(3):454–76.
10. Moberg E. Criticism and study of methods for examining sensibility in the hand. Neurology. 1962;12:8–19.
11. Dellon AL. The sensational contributions of Erik Moberg. J Hand Surg Br. 1990;15(1):14–24.
12. Johansson RS, Flanagan JR. Coding and use of tactile signals from the fingertips in object manipulation tasks. Nat Rev Neurosci. 2009;10(5):345–59.
13. Lundborg G. Nerve injury and repair. Regeneration, reconstruction and cortical remodelling. 2nd ed. Philadelphia: Elsevier; 2004.
14. Johansson RS, Landstrom U, Lundstrom R. Responses of mechanoreceptive afferent units in the glabrous skin of the human hand to sinusoidal skin displacements. Brain Res. 1982;244(1):17–25.
15. Vallbo AB, Johansson RS. Properties of cutaneous mechanoreceptors in the human hand related to touch sensation. Hum Neurobiol. 1984;3(1):3–14.
16. Edin BB. Quantitative analysis of static strain sensitivity in human mechanoreceptors from hairy skin. J Neurophysiol. 1992;67(5):1105–13.
17. Edin BB, Johansson N. Skin strain patterns provide kinaesthetic information to the human central nervous system. J Physiol. 1995;487(Pt 1):243–51.
18. Backlund H. Functional aspects of tactile directional sensibility. Gothenburg: University of Gothenburg; 2004.
19. Norrsell U, Backlund H, Gothner K. Directional sensibility of hairy skin and postural control. Exp Brain Res. 2001;141(1):101–9.
20. Sacks OW. A leg to stand on. Rev ed. London: Picador; 1991.
21. Maravita A, Iriki A. Tools for the body (schema). Trends Cogn Sci. 2004;8(2):79–86.

22. Johnson-Frey SH. What's so special about human tool use? Neuron. 2003;39(2):201–4.
23. Gentilucci M, Roy AC, Stefanini S. Grasping an object naturally or with a tool: are these tasks guided by a common motor representation? Exp Brain Res. 2004;157(4):496–506.
24. Brand P, Yancey P. Pain: the gift nobody wants. New York: Zondervan; 1993.
25. Sundelin A, Sörman A. Skammens hud: om spetälska i Sverige (in Swedish). Stockholm: Bokförlaget DN; 2004.
26. Larsson S. Girl who kicked the hornet's nest. Suffolk, London: MacLehose; 2009.
27. Minde JK. Norrbottnian congenital insensitivity to pain. Acta Orthop Suppl. 2006;77(321):2–32.
28. Minde J. Norrbottnian congenital insensitivity to pain. Umeå: Umeå University; 2006.
29. Minde J, Svensson O, Holmberg M, Solders G, Toolanen G. Orthopedic aspects of familial insensitivity to pain due to a novel nerve growth factor beta mutation. Acta Orthop. 2006;77(2):198–202.
30. Minde J, Toolanen G, Andersson T, Nennesmo I, Remahl IN, Svensson O, et al. Familial insensitivity to pain (HSAN V) and a mutation in the NGFB gene. A neurophysiological and pathological study. Muscle Nerve. 2004;30(6):752–60.
31. Einarsdottir E, Carlsson A, Minde J, Toolanen G, Svensson O, Solders G, et al. A mutation in the nerve growth factor beta gene (NGFB) causes loss of pain perception. Hum Mol Genet. 2004;13(8):799–805.

Chapter 8
The Sensational Brain

Abstract The exceptionally well-developed sensory functions of the hand are based on a delicate interaction between hand and brain. Tactile stimuli on the hand activate sensory receptors that are responsive to pressure, vibration, tension, cold, heat and nociceptive stimuli. Sensory signals reach the somatosensory cortex in the contralateral hemisphere via nerve pathways in the spinal cord. A cortical body map was defined more than 75 years ago showing that all parts of the body are represented within specific areas in somatosensory and motor cortices. The hand, face, mouth and tongue occupy the major parts, reflecting their delicate sensory and motor functions. Processing of tactile stimuli from the hand is an extremely complicated physiological process based on several parallel body maps processing information from various types of receptors in the skin as well as deeper tissue structures.

What are the secrets behind the refined motor and sensory properties of our hands? How can the sensibility in our hands be so well developed, and how can the hand and the brain interact so perfectly when we perform delicate precision tasks and when we perceive and recognise the roughness in the surface of a brick, the softness of silk and velvet or the shape and structure of a walnut?

When we touch an object or move our fingers over an irregular surface, small displacements of the skin of the fingertips create vibrations. In the hand, and especially in the hairless skin of the finger pulps, there are large numbers of sensory receptors (mechanoreceptors), responding to touch (pressure), vibrations or stretching (Fig. 8.1) [1–8]. There are also various types of free nerve endings that are activated by cold, heat or painful stimuli (nociceptive stimuli). Tactile stimulation of the sensory receptors of the fingers generates electric signals in large myelinated nerve fibres of the hand, and via nerve trunks in the hand and arm and pathways in the spinal cord, these sensory signals reach the somatosensory cortex, primarily in the contralateral hemisphere in the brain (Fig. 8.2). Thus, the sensory stimuli occur in the fingertips, but the perception and sensory experiences occur in the brain. We are dealing with an extremely complicated system of nervous pathways and relay

G. Lundborg, *The Hand and the Brain*,
DOI 10.1007/978-1-4471-5334-4_8, © Springer-Verlag London 2014

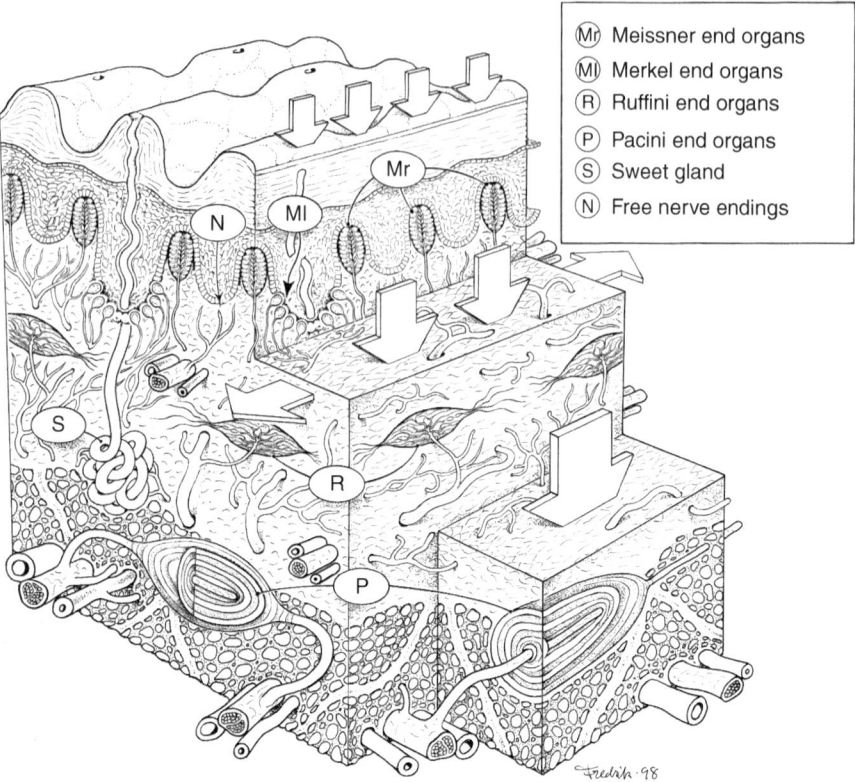

Mr Meissner end organs
MI Merkel end organs
R Ruffini end organs
P Pacini end organs
S Sweet gland
N Free nerve endings

Fig. 8.1 Superficial skin segment from the pulp of a finger with the ridges of the fingerprint visible at the top. Mechanoreceptors in the skin of the hand and fingers detect pressure (Merkel end organs), vibration (Meissner and Pacini end organs) and tension and stretching (Ruffini end organs) (Illustration: Fredrik Johansson. From Lundborg [2])

stations in the brainstem and in the thalamus before the signals reach the somatosensory cortex; much can happen along the route from the hand to the brain that might influence the final sensory experience [2].

The sensory mechanoreceptors in the hand interact to send appropriate sensory information to the brain when the hand is touched or when the hand is involved in an active 'act of touch' process where the sensory feedback is essential – for instance when we make meatballs, knead dough, test the consistency of bread or move our fingers across a surface to feel its texture. The sensory receptors that react to vibrations (Pacini end organs and Meissner end organs) are located in deep and superficial layers and are activated by vibrations within large and small cutaneous areas, respectively (Fig. 8.1). Merkel end organs adapt slowly and mainly respond to vibrations below 50 Hz, while Pacini end organs, which are fast adapting, mainly react to vibrations above 50 Hz. Merkel end organs, also called the Merkel disc receptors, located superficially at the centre of the papillary ridges of the skin, adapt slowly and react to static pressure; Ruffini end organs, found in deep dermal layers primarily on the dorsal aspect of the hand, are activated by tension and

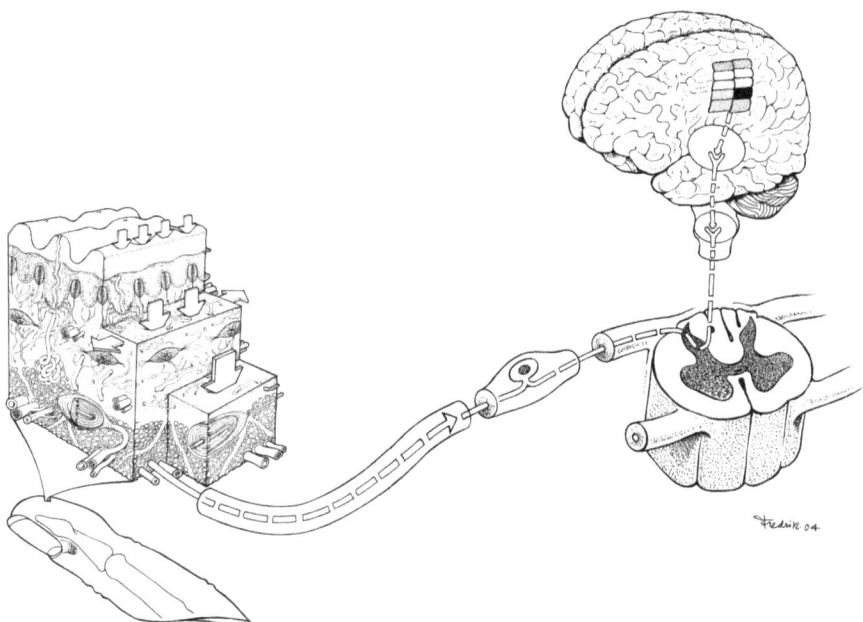

Fig. 8.2 The sensory pathway of sensory impulses induced by tactile stimulation of a finger. Afferent signals from cutaneous mechanoreceptors travel through large myelinated nerve fibres surrounded by a myelin sheath. They reach the somatosensory cortex after passing the dorsal root ganglia, up the dorsal column of the spinal cord via the medial meniscus pathway and the intermediate relay stations in the cuneate nucleus in the brainstem and the ventroposterior nucleus of the thalamus. Signals, elicited by touch, are primarily transmitted to the contralateral hemisphere

stretching of the skin [1–4]. Ruffini end organs in the skin on the back of the hand are subjected to tension when the hand is closed and they can provide the brain with important information about the position of the fingers based on their sensitivity to stretching [9, 10].

Georges Debrégeas and his research colleagues at the École Normale Supérieure in Paris have shown that the skin ridges comprising the fingerprints are important for optimising sensibility in the fingertips [11]. When the fingertips are moved across a surface, vibrations occur within frequencies that fit well with the most sensitive frequency band of Pacini end organ. The effect is most obvious when the fingertips move at a right angle against the orientation of the ridges. The skin ridges of the fingerprint are oval so that regardless of how the fingers are moved across a surface, there are always some ridges oriented at a right angle to the movement's direction. The findings indicate that fingerprint ridges play an important role in enhancing the hand's sensibility as well as making it easier to grip slippery objects, much like the treads of a tyre. Sweating – the sudomotor function – in the fingers is important for enhancing grip functions. Completely dry fingers have difficulty holding on to objects.

The nerve fibres in the hand that are activated by painful stimuli are small sized and not surrounded by a myelin sheath. They constitute an important warning system so that the hand reacts to the pricks of sharp items, heat and other factors that might harm it. A special system of free nerve endings is activated by heat; another system is activated by cold stimuli. A special system of small-sized unmyelinated fibres has been found to be associated with the pleasant feeling of a light touch. Stimulation of such tactile afferents in hairy skin activates the insular region of the brain without activating somatosensory cortex [12, 13]. These fibres may be activated upon skin contact between individuals during caresses and intimate touch.

Somatosensory Cortex

The somatosensory cortex has three major divisions: the primary (S I) and secondary (S II) somatosensory cortices and the posterior parietal cortex. In the primary somatosensory cortex, there is a delicate somatotopic organisation, a 'somatotopic map' or cortical body map so that sensory information reaching the brain from various parts of the body is effectively sorted out in a well-organised fashion (Fig. 8.3) [14–17]. Thus, sensory information from, for instance, the index finger arrives at a well-defined final destination that does not normally overlap areas receiving sensory information from the nearby middle finger (Fig. 8.4). However, this well-defined cortical mapping of individual fingers in the primary somatosensory cortex may be easily altered as a result of various events occurring in the hand, for instance a nerve injury or long-term exposure to vibration.

Sensory perception is an enormously complex function. A rich network of sensory association areas becomes involved in processing sensory information from the hand, and interpretation and understanding of the sensory information involves higher-level functioning brain areas, a more complex conscious mind. A large part of the human brain cortex consists of association areas, interacting with memories and emotional experiences as well as with other areas in brain cortex that receive input from other senses like vision and hearing. Thus, the sensory inflow from the hand is one of several important components that provide a complete and total experience together with input from several senses.

The sensory areas of the brain are also directly linked to the motor areas of the brain that control muscle activity. The primary motor cortex (M I) is situated immediately anterior to the central sulcus and primary somatosensory cortex and has a somatotopic organisation resembling that in the sensory cortex (Fig. 8.3). However, it seems that patterns of movements rather than separate muscles or body parts are somatotopically represented in the motor cortex [18, 19]. For instance, animal experiments have shown that stimulating one specific site may cause the mouth to open or cause the hand to form a grip and move to the mouth [20].

The cortical motor system is well integrated with the sensory cortical areas, and an afferent nerve inflow is necessary to perform well-coordinated movements [2, 21–26]. Disturbances in sensory functions of the hand can therefore negatively influence fine motor function as well as the grip force of the hand. Although executing motor movements involves the primary motor cortex (M1), other areas are

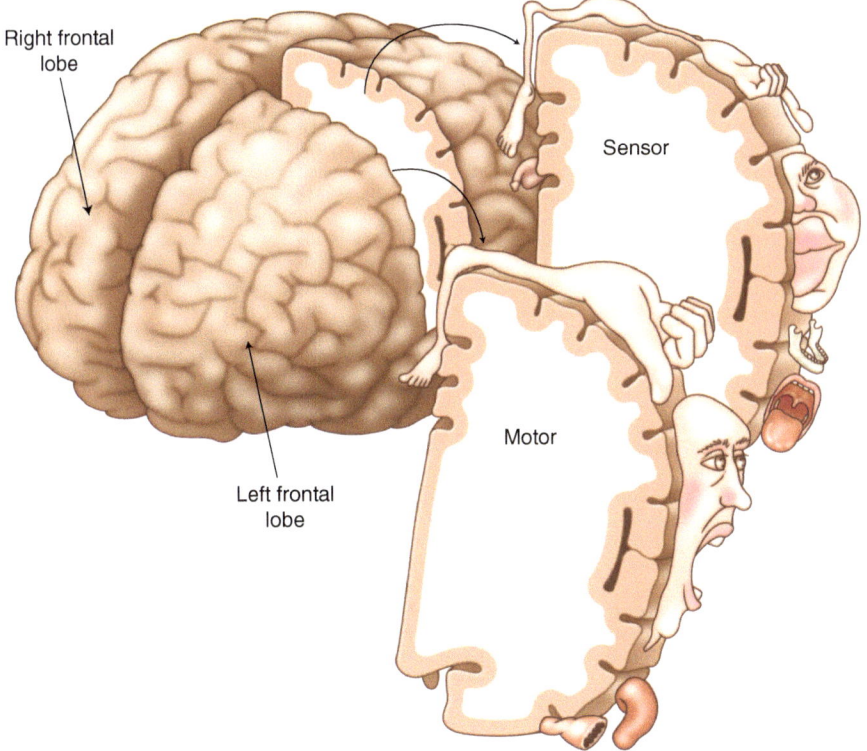

Fig. 8.3 The cortical body map as described by Penfield and Boldrey as early as 1937. The brain is seen obliquely from the frontal aspect, the arrow pointing forwards (anteriorly) and the areas for sensory and motor functions have been extracted as separate three-dimensional sections. The somatosensory cortical area is situated immediately posterior to the area for motor functions, the two cortical areas being separated by the central sulcus. The hand is represented in very large areas of the somatosensory and motor cortices. The hand representation is located close to the face. The thumb is especially big in this area and is located closest to the face

devoted to initiating and planning motor activities [27]. In brief, the premotor cortex, located immediately frontal to M1, prepares commands for voluntary actions which are then executed by the primary motor cortex [28].

Thus, the mechanisms behind the execution of muscle movements are complex and involve several brain areas. Desmurget investigated the effect of direct cortical stimulation of premotor and parietal regions in patients undergoing brain surgery for tumour removal [29]. When the premotor region of the frontal lobe was stimulated, complex multijoint movements were induced, but the patients did not feel that these movements were produced by a conscious internal act of will. However, stimulation of the parietal lobe provoked the conscious experience of wanting to move the upper limb, lips or tongue without any concomitant motor activity. When stimulation intensity was increased, patients believed that they had actually moved or talked, but again no muscle activity was detected.

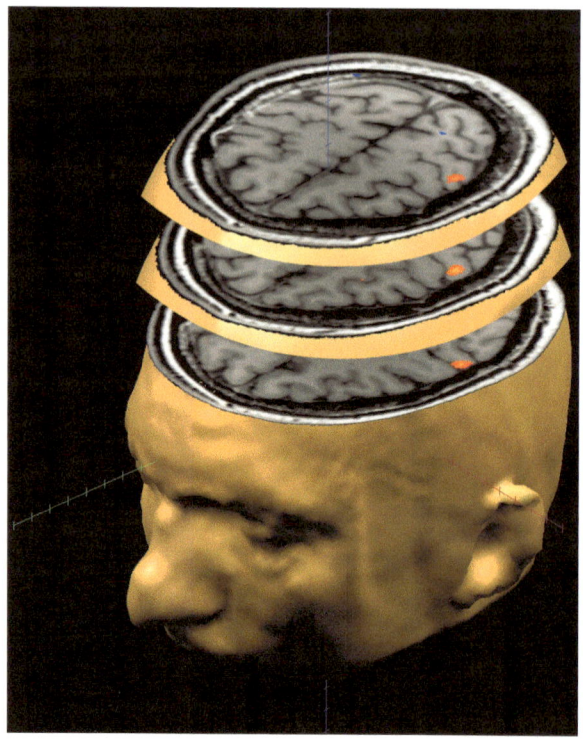

Fig. 8.4 Functional magnetic resonance imaging (fMRI) technology can be used to define the cortical representations of individual fingers in the somatosensory cortex. In this picture, the brain has been sliced in horizontal sections, and the representational areas for the right thumb (*bottom slice*), index finger (*middle slice*) and little finger (*upper slice*) have been activated by touching these fingers (Courtesy of Andreas Weibull, Medical Radiation Physics, Malmö, Skåne University Hospital, Malmö, Lund University)

The Cortical Hand Map

Seventy-five years ago, Wilder Penfield (1891–1976), a Canadian neurosurgeon and founder of the Montreal Neurological Institute, demonstrated how our various body parts are represented in the brain in a 'cortical body map' in both the sensory and motor cortices [14]. Penfield and Boldrey performed brain surgery on epileptics, exposing the surfaces of their brains after anaesthetising the skin in the surgical area. Thus, the patients were fully awake during surgery, and with an electric stimulator, Penfield could carefully stimulate various areas of the brain surface while asking the patient how and where the stimulus was felt. In this way, he could define a 'map' of the brain surface where the various parts of the body were represented. Electrical stimulation close to the upper side of the brain within the sensory cortex could be felt in the knee and thigh, while stimulation further down along the outside of the brain could be felt in the thoracic area and trunk. Further down the brain surface within the sensory cortex, he could identify areas corresponding to the hand and even further down there were areas representing the skin of the face, the mouth and the tongue. In analogy, stimulation of specific areas within the motor cortex, situated anteriorly to the sensory cortex, resulted in muscle contractions.

Penfield's original theories about the cortical body map have been verified using modern brain imaging technology, although some modifications have been suggested [15]. The locations of the individual fingers of the hand in the brain cortex

Fig. 8.5 In the homunculus figure, the body parts are proportioned to demonstrate how well they are represented in the brain. Natural History Museum, London

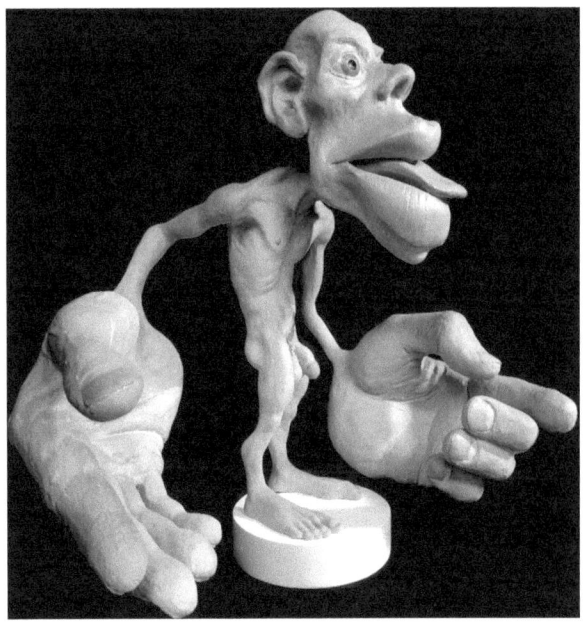

have been demonstrated in detail (Fig. 8.4) [15–17]. The sensory and motor cortices have much larger areas, i.e. more nerve cells, for the hand, face, mouth and tongue than, for instance, the foot, the lower extremities and the trunk (Fig. 8.3). The large representational areas correspond well to the enormously well-developed sensory and muscle functions of the hand, face and mouth, requiring a majority of the brain's resources. The hand and the mouth require more 'brain power' than other body parts.

Homunculus is the term for a symbolic figure whose body parts are presented in proportion to their representation in the brain – the more representation in the brain, the larger the body parts in the homunculus figure (Fig. 8.5). The homunculus figure is certainly not a beautiful one; it has grotesque proportions in its bizarre illustration of the relative amounts of brain power that are required to achieve sufficiently refined sensibility and motor functions in various parts of our body.

The cortical body map that was originally described by Penfield and Boldrey is simple and easy to understand, but today we know that there are several parallel body maps within the somatosensory cortex, and each map area receives and processes different types of sensory stimuli. There are four distinct somatotopic maps in the primary sensory cortex [30, 31]. These maps correspond well to four areas of S1 that have been identified by Brodman and named 1, 2, 3a and 3b [2]. The maps are parallel to each other, each of them representing one type of sensory information. Proprioceptive information from muscles and joints are represented in area 3a, whereas information from the skin, important for touch, is represented in area 3b. This information from the skin is further processed within area 1 and then combined with information from muscles and joints in area 2 [32].

The way the brain processes and interprets signals from the sensory receptors of the hand is complicated and not completely understood. For instance, sensory information from the hand related to micro-geometry (like textures and surface structures) and macro-geometry (shapes and sizes) are processes in different parts of the somatosensory cortex [33, 34]. It is well known which areas in the sensory cortex are activated when stimulation is applied to the sensory receptors of the fingers, but it is still a mystery how this electrical activity is converted in our consciousness into a subjective experience of shape or texture.

References

1. Johansson RS, Flanagan JR. Coding and use of tactile signals from the fingertips in object manipulation tasks. Nat Rev Neurosci. 2009;10(5):345–59.
2. Lundborg G. Nerve injury and repair. Regeneration, reconstruction and cortical remodelling. 2nd ed. Philadelphia: Elsevier; 2004.
3. Johansson RS, Landstrom U, Lundstrom R. Responses of mechanoreceptive afferent units in the glabrous skin of the human hand to sinusoidal skin displacements. Brain Res. 1982;244(1):17–25.
4. Vallbo AB, Johansson RS. Properties of cutaneous mechanoreceptors in the human hand related to touch sensation. Hum Neurobiol. 1984;3(1):3–14.
5. Johansson RS, Westling G. Roles of glabrous skin receptors and sensorimotor memory in automatic control of precision grip when lifting rougher or more slippery objects. Exp Brain Res. 1984;56(3):550–64.
6. Johansson RS, Vallbo AB. Tactile sensibility in the human hand: relative and absolute densities of four types of mechanoreceptive units in glabrous skin. J Physiol. 1979;286:283–300.
7. Johansson RS, Vallbo AB. Tactile sensory coding in the glabrous skin of the human hand. Trends Neurosci. 1983;6:27–32.
8. Stark B, Carlstedt T, Hallin RG, Risling M. Distribution of human Pacinian corpuscles in the hand. A cadaver study. J Hand Surg Br. 1998;23(3):370–2.
9. Edin BB. Quantitative analysis of static strain sensitivity in human mechanoreceptors from hairy skin. J Neurophysiol. 1992;67(5):1105–13.
10. Edin BB, Johansson N. Skin strain patterns provide kinaesthetic information to the human central nervous system. J Physiol. 1995;487(Pt 1):243–51.
11. Scheibert J, Leurent S, Prevost A, Debregeas G. The role of fingerprints in the coding of tactile information probed with a biomimetic sensor. Science. 2009;323(5920):1503–6.
12. Olausson H, Lamarre Y, Backlund H, Morin C, Wallin BG, Starck G, et al. Unmyelinated tactile afferents signal touch and project to insular cortex. Nat Neurosci. 2002;5(9):900–4.
13. Loken LS, Wessberg J, Morrison I, McGlone F, Olausson H. Coding of pleasant touch by unmyelinated afferents in humans. Nat Neurosci. 2009;12(5):547–8.
14. Penfield W, Boldrey E. Somatic motor and sensory representations in the cerebral cortex of man as studied by electrical stimulation. Brain. 1937;60:389–443.
15. Nakamura A, Yamada T, Goto A, Kato T, Ito K, Abe Y, et al. Somatosensory homunculus as drawn by MEG. Neuroimage. 1998;7(4 Pt 1):377–86.
16. van Westen D, Fransson P, Olsrud J, Rosen B, Lundborg G, Larsson EM. Fingersomatotopy in area 3b: an fMRI-study. BMC Neurosci. 2004;5:28.
17. Weibull A, Bjorkman A, Hall H, Rosen B, Lundborg G, Svensson J. Optimizing the mapping of finger areas in primary somatosensory cortex using functional MRI. Magn Reson Imaging. 2008;26(10):1342–51.

18. Schieber MH. Constraints on somatotopic organization in the primary motor cortex. J Neurophysiol. 2001;86(5):2125–43.
19. Schieber MH, Hibbard LS. How somatotopic is the motor cortex hand area? Science. 1993;261(5120):489–92.
20. Graziano MS, Taylor CS, Moore T. Complex movements evoked by microstimulation of precentral cortex. Neuron. 2002;34(5):841–51.
21. Roland PE, Zilles K. Functions and structures of the motor cortices in humans. Curr Opin Neurobiol. 1996;6(6):773–81.
22. Lindberg P, Forssberg H, Borg J. Rehabilitation after stroke. Imaging techniques show how the cortical reorganization is affected by training. Lakartidningen. 2003;100(51–52):4289–92.
23. Young JP, Geyer S, Grefkes C, Amunts K, Morosan P, Zilles K, et al. Regional cerebral blood flow correlations of somatosensory areas 3a, 3b, 1, and 2 in humans during rest: a PET and cytoarchitectural study. Hum Brain Mapp. 2003;19(3):183–96.
24. Lindberg C, Wunderlich M, Ratliff J, Dinsmore J, Jacoby DB. Regulated expression of the homeobox gene, rPtx2, in the developing rat. Brain Res Dev Brain Res. 1998;110(2):215–26.
25. Piitulainen H, Bourguignon M, De Tiege X, Hari R, Jousmaki V. Coherence between magnetoencephalography and hand-action-related acceleration, force, pressure, and electromyogram. Neuroimage. 2013;72:83–90.
26. Rosenkranz K, Rothwell JC. Modulation of proprioceptive integration in the motor cortex shapes human motor learning. J Neurosci. 2012;32(26):9000–6.
27. Haggard P. Neuroscience. The sources of human volition. Science. 2009;324(5928):731–3.
28. Tanji J, Mushiake H. Comparison of neuronal activity in the supplementary motor area and primary motor cortex. Brain Res Cogn Brain Res. 1996;3(2):143–50.
29. Desmurget M, Reilly KT, Richard N, Szathmari A, Mottolese C, Sirigu A. Movement intention after parietal cortex stimulation in humans. Science. 2009;324(5928):811–3.
30. Kandel ER, Schwartz JH, Jessell TM, Siegelbaum SA, Hudspeth AJ. Principles of neural science. 5th ed. New York: McGraw-Hill; 2013.
31. Kaas JH, Merzenich MM, Killackey HP. The reorganization of somatosensory cortex following peripheral nerve damage in adult and developing mammals. Annu Rev Neurosci. 1983;6: 325–56.
32. Gardner EP, Martin JH, Jessell TM. Coding of sensory information. In: Kandel ER, Schwartz JH, Jessell TM, editors. Principles of neural science. New York: McGraw-Hill; 2000. p. 411–29.
33. Bodegard A, Geyer S, Naito E, Zilles K, Roland PE. Somatosensory areas in man activated by moving stimuli: cytoarchitectonic mapping and PET. Neuroreport. 2000;11(1):187–91.
34. Bodegard A, Geyer S, Grefkes C, Zilles K, Roland PE. Hierarchical processing of tactile shape in the human brain. Neuron. 2001;31(2):317–28.

Chapter 9
How the Hand Shapes the Brain

Abstract The brain cortex contains more than 100 billion nerve cells and innumerable synaptic connections. The cortical body map is not fixed and hardwired, but can rapidly become reorganised as a result of a strengthening or weakening of the synaptic connections. The hand's representational area is experience dependent and can easily expand as a result of increased hand activity and increased sensory inflow. By contrast, decreased hand activity may result in a reduced hand representation. Anaesthetising the forearm skin enlarges the hand representational area, located immediately near the forearm representational area, and the result is rapid improvement of hand sensibility.

Nerve injuries may result in the misdirection of regenerating nerve fibres so that they reinnervate incorrect peripheral targets. As a result, the original hand representational area becomes completely reorganised into a mosaiclike pattern, a phenomenon that is associated with poor recovery of hand sensibility and severe persistent impairment of discriminative sensory capacity.

It was long believed that the cortical body map was fixed and hardwired from birth and could not be changed. But today another concept has evolved. We now know that the hand representation in the brain can rapidly change depending on how much, or how little, the hand is used and how much sensory input the brain receives from the hand. The same is true for all body parts; the brain is plastic and rapidly adaptable to change. Brain plasticity is a current key term when referring to learning, training and rehabilitation.

The brain cortex has more than 100 billion nerve cells and innumerable synaptic connections between nerve cells – neurons – forming networks of enormous dimensions. The functional plasticity of the brain is based on the fact that the function in such synaptic connections can be rapidly reinforced, weakened or inhibited. Existing but normally inhibited synapses may be rapidly awakened if the inhibition is withdrawn, resulting in very rapid alterations in synaptic function. The functions in the synaptic connections are constantly undergoing changes depending on changes in the activity of body parts, which was clearly demonstrated more than 50 years ago

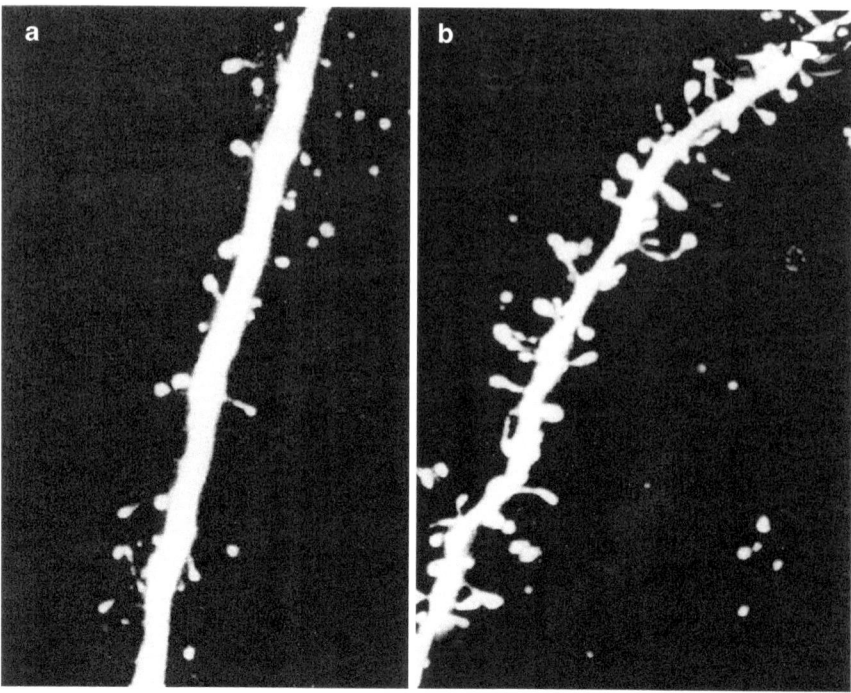

Fig. 9.1 High-resolution images of dendritic spines indicating synaptic connections from nerve fibres in the brains of (**a**) rats living in a standard environment and (**b**) rats living in an enriched environment. Rats living an active life with other rats in a stimulating environment exhibit many more active synapses in their brains than rats living alone in an activity-deprived poor environment without stimulation (Adapted from Johansson [6])

by Canadian neurophysiologist Donald Hebb [1–5]. Thus, the cortical representation of body parts is continuously modulated in response to activity, behaviour and skill acquisition. Much emphasis has been put on the neuronal dendritic spines, tiny protrusions from the long, slender dendritic extensions of the nerve cells, constituting the synaptic receptors. The appearance of such 'spines' constantly changes with activity and learning [6–8]. It is known from animal experiments that adult rats living in an enriched environment with other rats, and with access to various toys and activities, develop a more extensive network of synapses than those who live in solitude in a poor, activity-deprived environment [7, 8] (Fig. 9.1).

A very active hand results in increased activation of the nerve cells in the sensory and motor hand representational areas. As a consequence, the hand expands its representational area in the cortical hand map because it requires more brain resources, more 'brain space'. Thus, the hand can 'shape' the brain; the brain is functionally shaped based on the hand's experiences. If the hand, in contrast, is passive and immobile for a long time, its representation in the brain decreases and may totally disappear. Quite simply, the hand has to be active to maintain its representation in the brain – *use it or lose it*. On the other hand, we know that the hand representation in the brain can be re-established by training and activities.

Thus, the hand's representation in the brain is dependent on the hand's activities and experiences [9–13]. In advanced violinists, the left hand – 'the fingering hand' – has a larger than normal representation in the brain, especially if they start training early in Suzuki classes [14]. In addition, individual fingers can enlarge their representational cortical areas. If a finger is intensively subjected to tactile stimulation, it expands its representation in the sensory brain cortex [15, 16]. The fingers of blind patients, used to reading Braille, develop larger than normal representations in the sensory brain cortex [17, 18].

Increased use of the hand results in a larger representation in the brain, while, on the contrary, decreased use results in a reduced representational area that can result in impaired sensibility in the hand. Immobilisation in a splint for a few days results in a decreased capacity for precision movements, something that rapidly normalises as soon as the hand is remobilised [19]. If the hand is immobilised over an extended time period due to an injury, there may be reason to try to maintain its cortical representation using various training protocols. It is well known that the cortical hand representation can be activated by, for instance, observing active hands belonging to other people, imagining hand activities, reading about active hands or even by listening to 'action words' – words that describe hand activities (see Chap. 11). The same principles are useful in maintaining the cortical hand representation after the complete arrest of sensory input that occurs after a nerve transection in the hand or forearm – an important preparation for the sensory relearning programme that should be initiated when regenerating nerve fibres reach the hand after nerve repair (see below).

Confusion in the Cortical Hand Representation

The cortical hand map is maintained by normal use of the hands and the associated sensory input. This maintains the borders between the representational areas of separate fingers – a prerequisite for delicate, refined hand sensibility.

But if the sensory input from the hand differs from normal, it may affect the cortical hand representation. In people who work for long periods of time with hand-held vibrating tools, all parts of the hand are simultaneously exposed to vibrations – a type of long-term unphysiological repetitive sensory stimulation the brain does not recognise and that is difficult for the sensory cortex to process. The long-term consequence may be that the cortical hand map is reorganised towards a more irregular pattern where separate finger representational areas are enlarged and somewhat overlap [20–22]. Such changes can contribute to impaired sensibility and coordination of the hand.

Another analogous problem is *writer's cramp*, which is usually based on long periods of monotonous repetitive unphysiological hand movements. In such cases, the cortical hand map is reorganised and fine motor coordination is impaired. Another example along the same lines is *focal dystonia* – a condition that can be seen in musicians who practise extensively – and can result in difficulty controlling the movements in separate fingers (Fig. 9.2) [23–25].

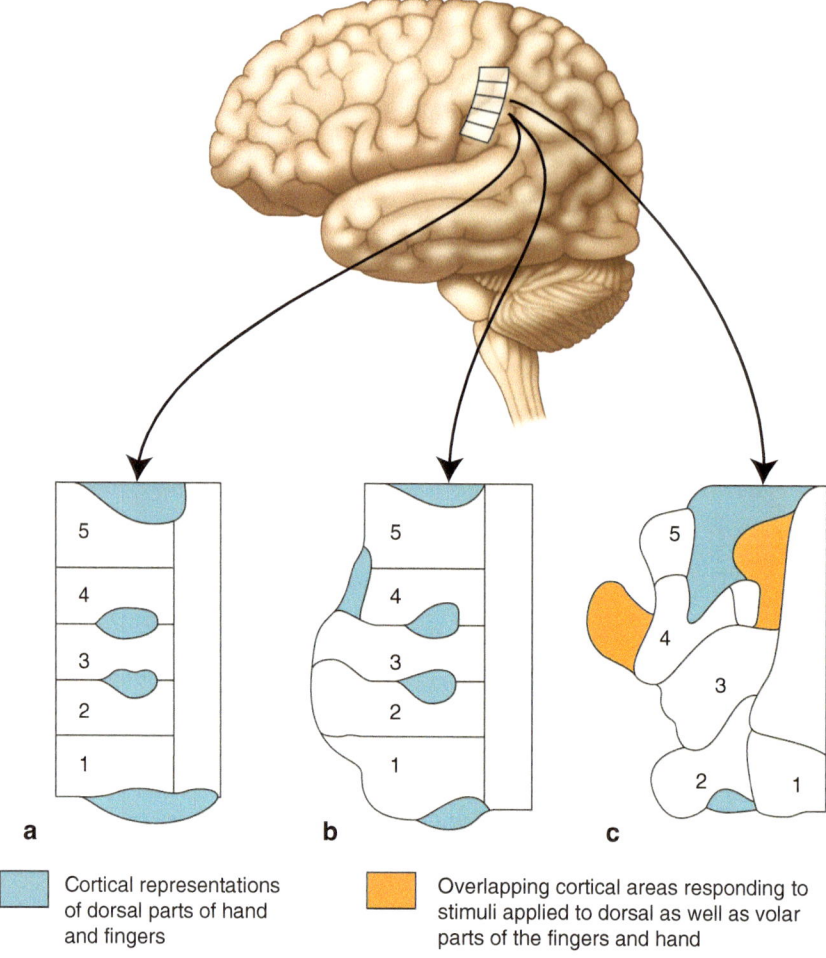

Cortical representations of dorsal parts of hand and fingers

Overlapping cortical areas responding to stimuli applied to dorsal as well as volar parts of the fingers and hand

Fig. 9.2 Normally the hand representation in the sensory cortex is well organised with obvious borders between individual finger representational areas (**A**). In this schematic drawing, a magnifying glass has been placed over the hand representation: *1* thumb, *2* index finger and so on. If the hand is subjected to too much sensory stimuli, the representation of the hand in the sensory brain cortex expands – 'the hand becomes larger in the brain'. If specific fingers are stimulated to a large extent, for instance the index and middle fingers, their representational areas increase (**B**). If the hand performs monotonous movements for a long time or is subjected to excessive sensory stimuli, its cortical representation changes and splits in a mosaiclike pattern (**C**), such as in focal dystonia, writer's cramp or vibration injury (Adapted from Lundborg [35])

Writer's cramp and focal dystonia are very difficult to treat; the problem is not primarily localised in the hand but rather in the brain. Specially designed training programmes are required to normalise the balance between hand and brain and to re-establish a normal cortical hand representation in the somatosensory cortex [26, 27].

Interacting Hemispheres

Most surgical procedures on the hand are performed under local anaesthesia of individual fingers or of the whole arm. Anaesthesia of individual fingers has a rapid and dramatic effect on their cortical hand representations: when the cortical representation of the anaesthetised finger suddenly becomes still due to lack of sensory input, the nearby representational areas of adjacent fingers expand so that they 'take over' the still area.

When the whole arm is anaesthetised by a so-called axillary nerve block, the corresponding cortical representational area is put into a state of vacancy due to loss of sensory input. This phenomenon has obvious effects on the sensibility of the contralateral non-anaesthetised arm. The gap between the two brain hemispheres is bridged by *corpus callosum*, which consists of nerve fibres connecting the left and right parts of the brain, allowing for a constant ongoing interaction between both hemispheres. If, let us say, the representational area of the right arm in the left hemisphere is put into a state of vacancy due to anaesthesia of the right arm, the non-anaesthetised left arm can benefit from this vacant area, gaining access to more brain space. The result is improved sensory function and a slightly increased grip force in the contralateral non-anaesthetised left hand [28–30].

How a Local Anaesthetic Cream, Applied to the Forearm, Can Improve Hand Sensibility

Knowledge about the cortical hand map and its capacity for rapid reorganisation has created exciting new possibilities for improving hand sensibility. If the skin of the forearm is anaesthetised with an anaesthetic cream, the corresponding representational area in the sensory cortex is put into a state of idleness due to a lack of sensory impulses. But there is a constant ongoing *cortical competition for space* in the brain among the various body parts, resulting in an interesting phenomenon when the forearm skin is anaesthetised. Following cutaneous anaesthesia of the forearm, the hand representational area, which is close to the inactive forearm representational area, gets a chance to expand through the activation of synaptic connections and will rapidly also incorporate the area that earlier represented the forearm, thus giving the hand access to more 'brain space' [31] (Fig. 9.3). A larger number of nerve cells than before are now involved in processing the sensory input from the hand. The result is that hand sensibility suddenly improves [32]. The effect can be maintained if the cutaneous anaesthetic procedure is repeated at a few days' interval and if it is repeated over a more extended time period at successively greater intervals, the effect may be long-lasting. This is a good illustration of the brain's capacity for adaptation to changes in its functional organisation and how we can utilise this capacity therapeutically in various types of sensory impairment in the hand [21, 29, 32–34].

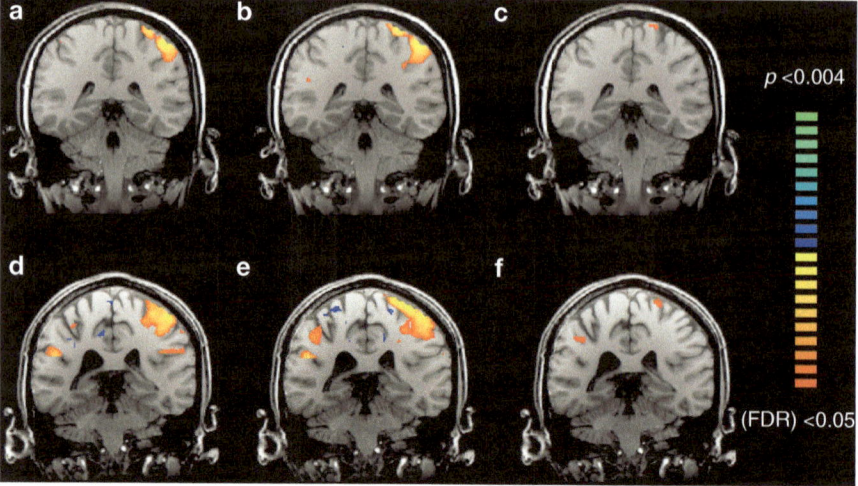

Fig. 9.3 Individual functional MR imaging results for two subjects – subject A (*upper row*, **a**–**c**) and subject B (*lower row*, **d**–**f**). Sensory stimulation of the right hand *before* application of an anaesthetic cream to the ipsilateral forearm resulted in a large activated area in the contralateral (left) somatosensory cortex S1 (**a**, **d**). After application of the anaesthetic cream to the right forearm, the contralateral S1 activation was expanded medially to the S1 hand area, corresponding to the forearm area (**b**, **e**). S1 reorganisation was confirmed by significant forearm activation, shown by comparing the sensory stimulation of the hand before and after the application of the anaesthetic cream (**c**, **f**). This investigation was conducted by Andreas Weibull, Medical Radiation Physics, Skåne University Hospital in Malmö (From Björkman et al. [31])

Nerve-Injured Hands

A cutting injury of the hand or forearm can easily result in the transection of the hand's most important sensory nerve, the median nerve which normally innervates the thumb, index finger, middle finger and half of the ring finger. This is a very serious injury, since the sensibility disappears in a large part of the hand, resulting in very poor hand function. Following surgical repair, new nerve fibres have to regenerate along the nerve pathways distal to the injury, a process that takes several months since nerve fibres grow at a maximum speed of 1 mm/day [35]. Frequently, regenerating nerve fibres advance into incorrect pathways so that, for instance, the nerve fibres of the thumb regenerate instead into the index finger and vice versa. From the brain's viewpoint, this is extremely problematic since the original representational area of the hand in the somatosensory cortex becomes completely reorganised into a chaotic mosaiclike pattern (Fig. 9.4) [35]. When this happens, a touch of the index finger can be interpreted by the brain as a touch of the thumb or vice versa. The hand is 'speaking a new language' to the brain.

The patient now has to learn to understand the new language and reacquire the capacity to recognise and perceive the shapes and textures of items by touching them – like a newborn child touching objects around it for the first time. Tactile and

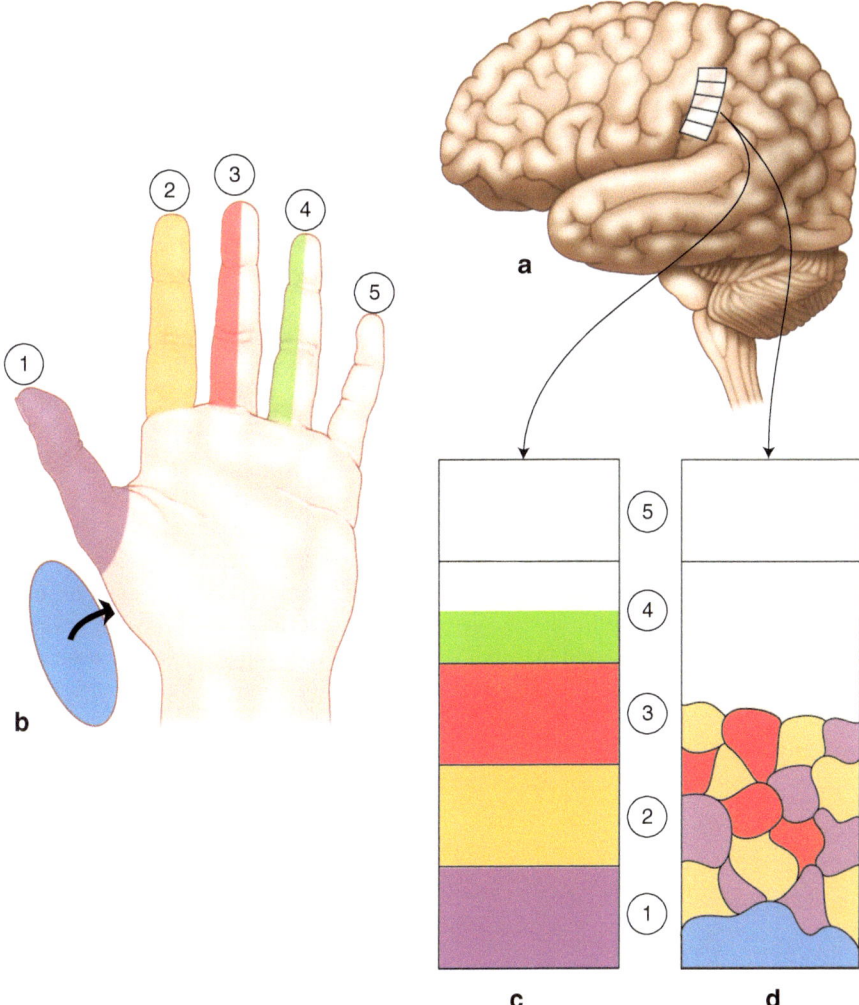

Fig. 9.4 Illustrations from animal experiments demonstrating the normal finger representational areas in the sensory brain cortex (**A–C**). The borders between the individual finger representational areas are obvious (**C**). After injury and repair, nerve fibres regenerate to the hand, but a large number of fibres advance into 'non-correct' fingers. The nerve fibres of the thumb may regenerate into the index finger and vice versa. This leads to extensive changes in the cortical hand map; the former regular pattern is changed into a mosaiclike pattern (**D**). 'The hand is speaking a new language to the brain'

visual impressions interact in newborns so that the child acquires a capacity to recognise and understand shapes by touching them, perhaps also using its lips to further optimise sensory information. An adult nerve-injured patient, due to reinnervation of incorrect cutaneous areas of the hand, usually loses the ability that he acquired as a newborn child to comprehend shapes and textures by touching them. Therefore, after nerve repair, adult patients must participate in specific sensory re-education programmes to recover as much as possible of their hand's tactile discriminative functions.

In the sensory relearning programme, the patient explores the shape and texture of objects of increasing 'difficulty' by touching them – alternately with the eyes open and closed. The visual sense *explains* what the hand just touched, thus recoding for the sense of touch [34–38]. By training on objects of continuously increasing complexity, it is possible to train the brain's capacity to understand shapes and textures, but an adult patient usually never recovers the original refined tactile sensibility in the hand after repair of an injured nerve.

While adults have difficulties learning the new sensory language after nerve repair, young children behave quite differently. Children up to 10–12 years of age usually have a complete restoration of all sensory functions in the hand including the fine discriminative capacities. They learn to understand the new 'sensory language', just as children moving to a foreign country can easily learn to speak and understand a second language [39].

There is an obvious analogy between the capacity to recover normal sensibility in the hand after nerve repair and the capacity to acquire a second language with correct grammar and pronunciation, as related to age [39]. In both situations, there seems to be a critical limit at the age of around 10–12 years, followed by a rapid decline during the teens and the adult stage. Interestingly, it has recently been demonstrated that the excellent sensory recovery that is usually obtained after nerve repair in children below the age of 12 persists even at a follow-up 30 years later when the child has reached adulthood [40, 41], much like the lifelong ability to speak a second language after learning it at a very young age.

References

1. Hebb DO. The effects of early experience on problem solving at maturity. Am J Psychol. 1947;2:737–45.
2. Hebb DO. The organization of behavior: a neuropsychological theory. New York: Wiley; 1949.
3. Kaas JH. Plasticity of sensory and motor maps in adult mammals. Annu Rev Neurosci. 1991;14:137–67.
4. Chen R, Cohen LG, Hallett M. Nervous system reorganization following injury. Neuroscience. 2002;111(4):761–73.
5. Wall JT, Xu J, Wang X. Human brain plasticity: an emerging view of the multiple substrates and mechanisms that cause cortical changes and related sensory dysfunctions after injuries of sensory inputs from the body. Brain Res Brain Res Rev. 2002;39(2–3):181–215.
6. Johansson BB. Brain plasticity and stroke rehabilitation. The Willis lecture. Stroke. 2000;31(1):223–30.
7. Trachtenberg JT, Chen BE, Knott GW, Feng G, Sanes JR, Welker E, et al. Long-term in vivo imaging of experience-dependent synaptic plasticity in adult cortex. Nature. 2002;420(6917): 788–94.
8. Johansson BB, Belichenko PV. Neuronal plasticity and dendritic spines: effect of environmental enrichment on intact and postischemic rat brain. J Cereb Blood Flow Metab. 2002; 22(1):89–96.

9. Pascual-Leone A. The brain that plays music and is changed by it. Ann N Y Acad Sci. 2001;930:315–29.
10. Pascual-Leone A, Hamilton R. The metamodal organization of the brain. Prog Brain Res. 2001;134:427–45.
11. Rauschecker JP. Cortical plasticity and music. Ann N Y Acad Sci. 2001;930:330–6.
12. Lundborg G. Brain plasticity and hand surgery: an overview. J Hand Surg Br. 2000; 25(3):242–52.
13. Lundborg G, Richard P. Bunge memorial lecture. Nerve injury and repair–a challenge to the plastic brain. J Peripher Nerv Syst. 2003;8(4):209–26.
14. Elbert T, Pantev C, Wienbruch C, Rockstroh B, Taub E. Increased cortical representation of the fingers of the left hand in string players. Science. 1995;270(5234):305–7.
15. Merzenich MM, Jenkins WM. Reorganization of cortical representations of the hand following alterations of skin inputs induced by nerve injury, skin island transfers, and experience. J Hand Ther. 1993;6(2):89–104.
16. Merzenich MM, Kaas JH, Wall JT, Sur M, Nelson RJ, Felleman DJ. Progression of change following median nerve section in the cortical representation of the hand in areas 3b and 1 in adult owl and squirrel monkeys. Neuroscience. 1983;10(3):639–65.
17. Pascual-Leone A, Torres F. Plasticity of the sensorimotor cortex representation of the reading finger in Braille readers. Brain. 1993;116(Pt 1):39–52.
18. Pascual-Leone A, Wassermann EM, Sadato N, Hallett M. The role of reading activity on the modulation of motor cortical outputs to the reading hand in Braille readers. Ann Neurol. 1995;38(6):910–5.
19. Lissek S, Wilimzig C, Stude P, Pleger B, Kalisch T, Maier C, et al. Immobilization impairs tactile perception and shrinks somatosensory cortical maps. Curr Biol. 2009;19(10):837–42.
20. Lundborg G, Rosen B, Knutsson L, Holtas S, Stahlberg F, Larsson EM. Hand-arm-vibration syndrome (HAVS): is there a central nervous component? An fMRI study. J Hand Surg Br. 2002;27(6):514–9.
21. Rosen B, Bjorkman A, Lundborg G. Improved hand function in a dental hygienist with neuropathy induced by vibration and compression: the effect of cutaneous anaesthetic treatment of the forearm. Scand J Plast Reconstr Surg Hand Surg. 2008;42(1):51–3.
22. Bjorkman A, Weibull A, Svensson J, Balogh I, Rosen B. Cortical changes in dental technicians exposed to vibrating tools. Neuroreport. 2010;21(10):722–6.
23. Byl NN, Melnick M. The neural consequences of repetition: clinical implications of a learning hypothesis. J Hand Ther. 1997;10(2):160–74.
24. Elbert T, Candia V, Altenmuller E, Rau H, Sterr A, Rockstroh B, et al. Alteration of digital representations in somatosensory cortex in focal hand dystonia. Neuroreport. 1998;9(16): 3571–5.
25. Altenmuller E. Focal dystonia: advances in brain imaging and understanding of fine motor control in musicians. Hand Clin. 2003;19(3):523–38, xi.
26. Byl NN, McKenzie A. Treatment effectiveness for patients with a history of repetitive hand use and focal hand dystonia: a planned, prospective follow-up study. J Hand Ther. 2000;13(4):289–301.
27. Byl NN, Nagarajan SS, Merzenich MM, Roberts T, McKenzie A. Correlation of clinical neuromusculoskeletal and central somatosensory performance: variability in controls and patients with severe and mild focal hand dystonia. Neural Plast. 2002;9(3):177–203.
28. Bjorkman A, Rosen B, van Westen D, Larsson EM, Lundborg G. Acute improvement of contralateral hand function after deafferentation. Neuroreport. 2004;15(12):1861–5.
29. Bjorkman A, Rosen B, Lundborg G. Enhanced function in nerve-injured hands after contralateral deafferentation. Neuroreport. 2005;16(5):517–9.
30. Bjorkman A, Rosen B, Lundborg G. Anaesthesia of the axillary plexus induces rapid improvement of sensory function in the contralateral hand: an effect of interhemispheric plasticity. Scand J Plast Reconstr Surg Hand Surg. 2005;39(4):234–7.
31. Bjorkman A, Weibull A, Rosen B, Svensson J, Lundborg G. Rapid cortical reorganisation and improved sensitivity of the hand following cutaneous anaesthesia of the forearm. Eur J Neurosci. 2009;29(4):837–44.

32. Bjorkman A, Rosen B, Lundborg G. Acute improvement of hand sensibility after selective ipsilateral cutaneous forearm anaesthesia. Eur J Neurosci. 2004;20(10):2733–6.

33. Rosen B, Bjorkman A, Lundborg G. Improved sensory relearning after nerve repair induced by selective temporary anaesthesia - a new concept in hand rehabilitation. J Hand Surg Br. 2006;31(2):126–32.

34. Lundborg G, Bjorkman A, Rosen B. Enhanced sensory relearning after nerve repair by using repeated forearm anaesthesia: aspects on time dynamics of treatment. Acta Neurochir Suppl. 2007;100:121–6.

35. Lundborg G. Nerve injury and repair. Regeneration, reconstruction and cortical remodelling. 2nd ed. Philadelphia: Elsevier/Churchill Livingstone; 2004.

36. Dellon AL. Evaluation of sensibility and re-education of sensation in the hand. Williams & Wilkins, cop: Baltimore; 1981.

37. Rosen B, Balkenius C, Lundborg G. Sensory re-education today and tomorrow. Review of evolving concepts. Br J Hand Ther. 2003;8:48–56.

38. Lundborg G, Rosen B. Hand function after nerve repair. Acta Physiol (Oxf). 2007;189(2): 207–17.

39. Lundborg G, Rosen B. Sensory relearning after nerve repair. Lancet. 2001;358(9284): 809–10.

40. Chemnitz A, Bjorkman A, Dahlin LB, Rosen B. Functional outcome 30 years after median and ulnar nerve repair in childhood and adolescence. J Bone Joint Surg Am. 2013;95(4):329–37.

41. Chemnitz A, Andersson G, Rosen B, Dahlin LB, Bjorkman A. Poor electroneurography but excellent hand function 31 years after nerve repair in childhood. Neuroreport. 2013; 24(1):6–9.

Chapter 10
The Interaction of the Senses

Abstract The sense of touch is essential for exploring the surrounding world, as the refined sensibility of the hand can identify shapes, forms and structures. However, all of our senses work together to create a full picture of the surrounding environment. The sense of touch works with vision, hearing, smell and taste to generate an inner picture of the outer world. If one sense is impaired or affected by illness, the other senses compensate by refining their capacities. Blind people reading in Braille develop improved sensibility in their fingers, and the vacant cortical visual areas become involved in processing the sensory stimuli from the hand. Deaf people can enjoy music by placing their hands on the speakers; the vibrations induced by the music activate the sensory cortex and cortical areas devoted to hearing. For patients lacking hand sensibility due to nerve injury, a Sensor Glove, equipped with miniature microphones in the fingers, can pick up the friction sounds that are induced when the hand moves across a surface, and the corresponding stimuli that activate auditory cortical areas also activate sensory cortical areas – the patient can listen to what the hand feels. In a clinical experimental setting, sensibility and vision can be co-activated in such a way that a person's hand sensibility can be transposed into a rubber hand or some other foreign object. Thus, the multisensory and multimodal capacities of the brain are key factors for experiencing and understanding the surrounding world, normally and in specific situations of sensory deprivation.

Our senses interact and work together to generate an inner picture of the outer world. The English physician and mystic Robert Fludd (1574–1637) argued that in man's mind sensory inputs from the outer world are combined with a divine inspiration to generate an inner picture of the external environment, a concept he illustrated in a clever image where the sensibility of the hand is ranked in the same category as vision and hearing, and where the hand, like the eye and ear, is described as a sense organ (Fig. 10.1). The interaction of our senses becomes especially obvious if one sense is temporarily or permanently defective or absent. For example, groping about to make your way in a dark cellar without using vision leads to

G. Lundborg, *The Hand and the Brain*,
DOI 10.1007/978-1-4471-5334-4_10, © Springer-Verlag London 2014

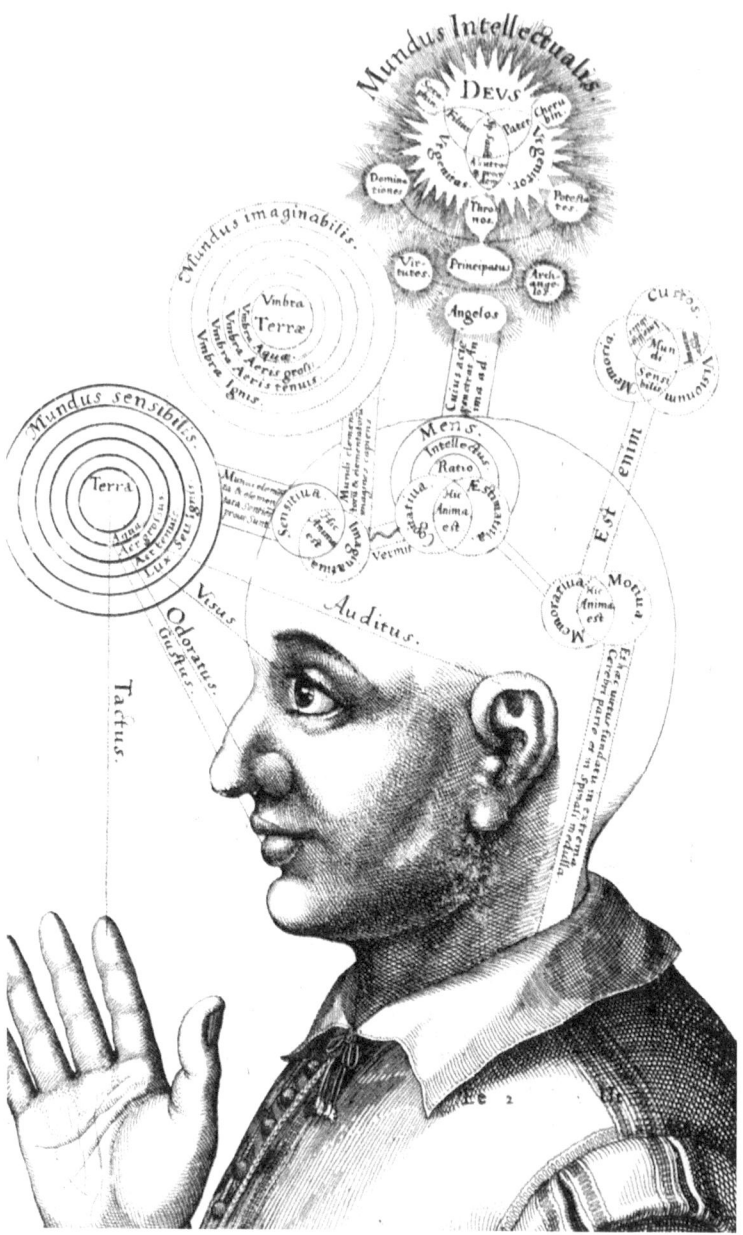

Fig. 10.1 Our senses work together to provide us with a complete picture of the surrounding world. The English physician and mystic Robert Fludd (1574–1637) argued that the sensory perceptions are combined with a divine inspiration to generate an inner picture of the world around us. Copperplate from Robert Fludd's *Tomus secundus de supernaturali, naturali, praeternaturali et contranaturali microcosmi historia* (Oppenheim, 1629) (From the exhibition Tankens Bilder (*Pictures of the Mind*), courtesy of Bildmuseét, Umeå, Sweden)

the remaining senses being sharpened; the sound of your footsteps and the smell of moisture and mould cooperate with your sensitive hands touching the walls to generate an inner picture of the non-visible cellar space.

Our lives become rich and full, thanks to the cooperation and interaction of our senses. Together vision, hearing, tactile sensibility, taste and smell generate a symphony of sensory impressions that enhances and amplifies the total experience of, for instance, a meal. Smells, colours and the appearance of the table setting are all important factors for stimulating the taste buds and the appetite. A fine wine tastes best in a beautifully shaped transparent glass that reveals its liveliness and colour. Who would enjoy champagne in a toothbrush glass or a tin mug? The taste and aroma of an orange are strongest when it is peeled so that the spray from its juice hits the skin and the smells generate an aura around the fruit. Crisps have to be crispy and salty; the hearing input complements the taste impressions. One element of training and re-educating the sensory functions of the hand after a nerve injury can be 'tactile meals'. In this type of training programme, several senses work together to create an assembled sensory impression and perception of the item that is touched. In such a tactile meal, the sensory input from the hand is combined with vision, hearing, smell and taste. The meal may include soft sugar-coated jelly sweets, rough pieces of liquorice, crisps and jellybeans, or the patient may be peeling oranges, cracking nuts and peeling shrimps.

Certainly our senses interact when we, for instance, walk through the woods on a rainy and stormy day in autumn; the moisture against our skin, the rustling of the leaves, the smells of the soil, the colours. Or, when packing a snowball – the feeling of newly fallen snow, the icy coldness, the packing resistance and the crunching sound of the snow being compressed.

The Multisensory Brain

The interaction among our senses is an expression of our brain's way of working, processing and coordinating sensory impressions from several senses at the same time [1]. According to classic concepts, there are well-defined cortical areas devoted to receiving and processing various types of sensory input: one area in the occipital lobe dedicated to processing visual input, another area in the temporal lobe for processing auditory input and a third area in the parietal lobe for receiving and processing tactile input from various parts of the body – the somatosensory cortex (Fig. 10.2). But this is a simplified description of how the brain receives and processes information from various senses. Our senses interact and cooperate with each other when interpreting sensory impressions via extensive cortical association areas, including earlier experiences. Within these areas, visual, hearing and tactile impressions cooperate to achieve a total experience of the surroundings and to solve specific problems. One example is estimation of distance, something that may require cooperation among vision, hearing and perhaps touch [2, 3].

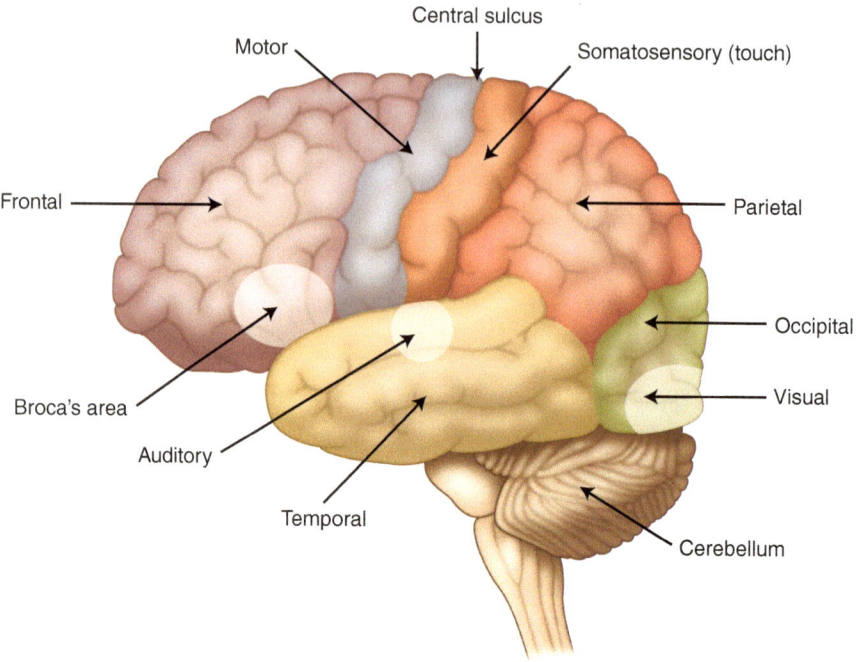

Fig. 10.2 Schematic drawing illustrating the lobes of the brain. According to classic concepts, various cortical areas are devoted to receiving and processing various types of sensory input: one area in the occipital lobe dedicated to processing visual input, another area in the temporal lobe for processing auditory input and a third area in the parietal lobe for receiving and processing tactile input from various parts of the body – the somatosensory cortex. However, this is a simplified description of how the brain receives and processes information from various senses, since senses interact and cooperate with each other when interpreting sensory impressions via extensive cortical association areas. Broca's area is important for speech and language ability. The somatosensory and motor cortices are localised behind and anterior to the central sulcus, respectively

Thus, the various sense areas in the brain act together and are closely connected to each other by complex systems of pathways and innumerable synapses. Some areas contain multisensory nerve cells that are activated by input from not only *one* type of sense, but they can be activated by two or more types of sensory input. Some of those nerve cells are activated by visual as well as tactile sensory input in a visuo-tactile interaction; others are activated by auditory as well as tactile sensory input in an audio-tactile interaction.

Sight and Sensation

The senses of touch and vision are normally closely connected. If you are going to move an apple from one plate to another, you first fix your eyes on the apple until your hand grasps it. Then you shift your eyes to the next plate and keep them there until your hand has put the apple in this new location. During these movements, the eyes are never shifted to the hand or the apple; you intuitively shift your vision to search for landmarks and then keep your eyes there [4]. With each movement the hand adapts the gripping force around the apple to avoid dropping it, while the arm's lifting power is automatically adjusted to move the apple from one plate to another.

The cooperation between vision and the sense of touch becomes even more evident when one of the senses is absent or does not function appropriately. A person wearing thick gloves loses much of the sensibility of the hand: it then becomes natural to rely on vision for a more effective sensory feedback from the hand. The same is true for hand amputees who use a hand prosthesis. The hand prostheses available today have no conscious sensibility, a fact that makes it important for the amputee to use vision instead to control and regulate the grip and movements of the prosthetic hand.

When one sense is absent, the brain adapts to the situation so that the corresponding 'vacant' projectional area takes on new functions and can thus provide more brain power to the remaining senses. This is true for humans as well as animals: in mice that are born blind, the sensibility of the whiskers increases. The whiskers' projectional area in the brain expands and will include the 'vacant' sensory cortical areas. Such a cortical reorganisation even occurs in healthy mice that live for a long time in total darkness [1].

In humans, loss of vision can lead to improved hand sensibility. The refined hand sensibility in blind people has been put to special use in health care in Germany. The gynaecologist Frank Hoffmann from Duisburg got the idea to train visually impaired women to palpate the breasts of women at risk of cancer. He started the project 'discovering hands' in 2006. The goal was to develop a cost-effective examination method while simultaneously creating a new work field for blind women. According to the innovators of the project, the extremely well-developed tactile sensibility in the blind women's hands allowed them to identify significantly smaller tumours in the breast tissues as compared to physicians with intact vision. The 'discovering hands' project was concluded at the end of 2008 when the Medical Society in Nordrhein examined the course participants and licensed them as the first practitioners of 'medical palpation' in Germany.

In the performing arts, the improved hand sensibility in blind people has been utilised in some theatres to enhance their experience of the theatre's play. Blind theatregoers were given the opportunity to meet and talk to the actors, to feel the contours of their faces and the character of their clothes. In this way they got to know the actors and develop close relationships with them, which enhanced the theatregoers' experience of the performance.

Reading Hands

I am blind.... What would I do to be able to see?
How can it be possible for me to read what has been written by a person with intact vision?
About history? About art? About medicine? About politics?
About women and men? About myself? About the mysteries of birth
and love? In short, how will it be possible for me, a
blind man, to find my place in the world as a part of the world?

Fig. 10.3 (**a**, **b**) Blind person reading Braille. The Braille alphabet is based on cells consisting of six raised dots and the presence or absence of dots that, in various combinations, provide a 64-character alphabet (Photo: Francisco Zizola (**a**), Jimmy Wahlstedt (**b**))

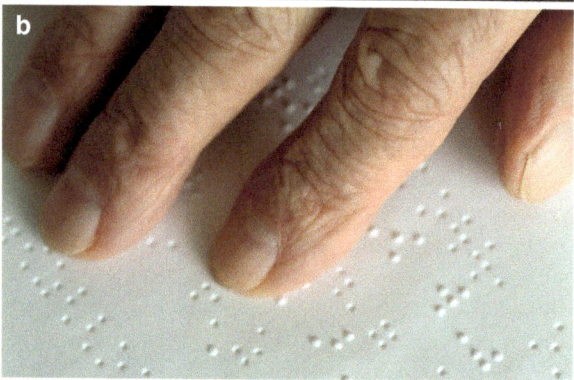

   The words are from a diary note of the French teacher Louis Braille, the inventor of the modern Braille alphabet. Louis Braille, born in 1809, was blinded as a result of an infected eye injury in childhood. Braille developed military systems for reading in darkness that became the first useful Braille alphabet, published in 1829. The concept was to utilise the delicate tactile sensibility of the finger pulps to compensate for missing vision. Braille was influenced by an earlier system created for blind people, developed by the French artillery captain Charles Barbier (1767–1841), which was based on 12 raised dots in each sign. The Barbier system was of phonetic nature based on phonetic sounds in the French language.

   The Braille alphabet is based on cells consisting of six raised dots, where the presence or absence of dots in various combinations provide a 64-character alphabet (Fig. 10.3). Braille finalised his system in 1825, but the Braille alphabet did not come into worldwide use until the 1860s. Unfortunately for Braille, who died of tuberculosis 1852, he never was able to experience this success. Today the Braille alphabet is in use worldwide; his system has never been surpassed. The fingers moving over the dots read the text in the same way that the eyes of sighted people scan a text.

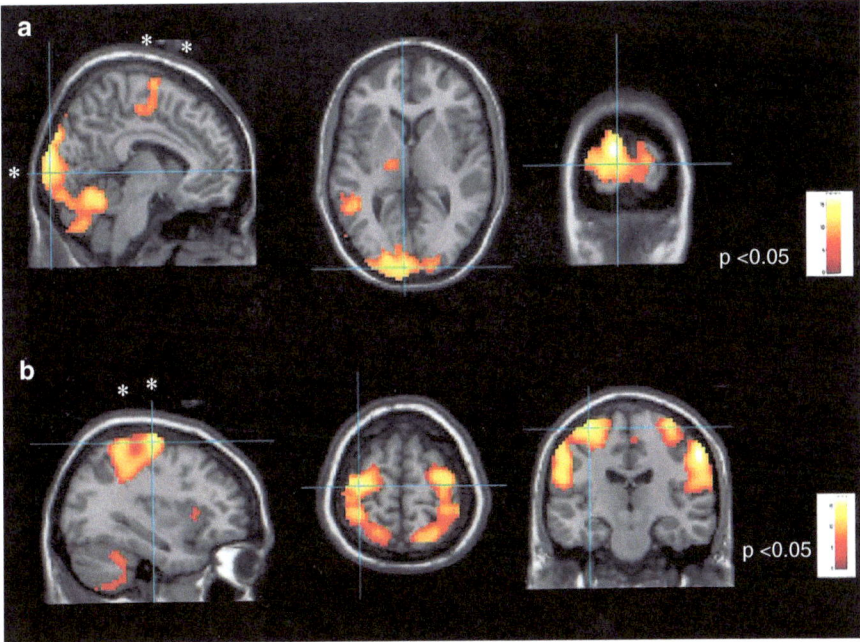

Fig 10.4 (**a**) When blind subjects read Braille with their right hand, it activates the visual cortex (*) as well as the sensorimotor areas (**). The visual area is 'vacant', making it an extra resource to support the sensory cortical areas in interpreting the tactile input from the hand as it reads. (**b**) When normal-sighted subjects move their right hand over the Braille alphabet, the results show the expected activation of sensorimotor areas but no activation of the visual cortex. The pictures are based on an fMRI investigation (From Gizewski et al. [6])

Several scientists have demonstrated that those who are born blind or acquired blindness in early childhood utilise the 'vacant' cortical visual area in the occipital lobe when they read the Braille alphabet with their fingers [5, 6]. When healthy people with intact vision move their finger pulps over the dots in the Braille alphabet, only the sensory cortex is activated; it is a tactile experience only, not a 'visual reading experience'. By contrast, in a blind person, the sensory and visual cortices work together to generate a reading experience transmitted via the finger pulps (Fig. 10.4). Both the sensory and visual cortices are activated when the fingers move over the dots. This means that more nerve cells are involved in the sensory experience, and more brain resources contribute to a refinement of the tactile sensibility of the fingers.

Fingers can read both text and pictures. In tactile picture books, the pictures consist of elevated shapes in various materials with varying textures. Such books are available at different levels of difficulty and for various ages. Picture books are very important in the development of all children, and for visually impaired children, tactile picture books are important for learning and understanding the meaning and symbolism of pictures.

Fig. 10.5 A deaf girl listens to orchestra music by holding her hands around an inflated balloon (From *Nature*, October 21, 2004. Courtesy of McMillan Publishers Ltd (*Nature*))

Listening Hands

There are examples on how the sensibility of hands can substitute for lost hearing. Tactile sensibility and hearing have much in common. In both cases, you are perceiving vibrations in the skin via sensory receptors responding to vibrations; in the ear via a vibrating tympanic membrane transmitting vibrations of various frequencies to hair cells situated in the cochlea of the inner ear. Therefore, it is natural that, in some situations, the sensory functions of the hand can substitute for lost hearing. However, the reverse can also be true: in some situations the hearing cells can replace absent sensibility in the hand.

One of my music-loving patients gradually lost her hearing due to illness and could no longer listen to the music that she had known for several years. She found a solution: by placing her hands on top of her stereo loudspeakers, she could feel the vibrations of the music and in that way enjoy it: the vibration sense of the hand substituted for the lost hearing sense; her 'listening' hands gave her back the ability to enjoy music. This was the case until 1 day she was affected by impaired sensibility in the hands due to bilateral carpal tunnel syndrome – a chronic compression of the most important sensory nerve, the median nerve: 'Doctor, I am about to become deaf', she said. But she was easily cured. A minor surgical procedure released the compressed sensory nerve, restoring her 'hearing', and once again she could enjoy music via the 'listening' hands.

Listening to music using the sensory capacities of the hands is probably a training process where the hearing areas in the temporal lobes are reorganised to instead process tactile sensory input from the hand – the vibrations generated by music. Figure 10.5 demonstrates such a situation in which a deaf girl in a concert hall listens to the music by holding an inflated balloon with both hands. The sound from the symphony orchestra on the stage induces vibrations in the balloon's membrane

Fig. 10.6 The Sensor Glove. Miniature microphones are incorporated on top of the fingers. Various materials induce different 'friction sounds'. Touching a Velcro strip doesn't sound the same as touching foam rubber or a piece of sheepskin. In this way a person with impaired sensibility of the hand can 'listen to what the hand feels' (Photo: Jimmy Wahlstedt)

that are transformed to her hands constituting a 'tactile tympanic membrane'. The girl listens to the music using her hands.

The Sensor Glove

When the hand is moved across a surface, it generates a slight friction sound that reflects the character and texture of the underlying surface – wood 'sounds' different than metal and glass sounds different than paper. Normally we do not pay attention to this, but if sensibility is lacking in the hand, such friction sounds may be important and may even offer a possibility for 'artificial sensibility' in the hand. The 'Sensor Glove' has miniature microphones placed on top of the fingertips that can register the faint friction sounds that are generated when the hand touches or moves over a surface (Fig. 10.6) [7–11]. The sound is transmitted to earphones in both ears. Using a sound processor, the sound is distributed so that sound from the thumb is mostly directed to the left ear and sound from the little finger is directed mostly to the right ear. In this way the hand gains a stereophonic sensibility based on sound perception and can rapidly be trained to localise touch within various parts of the hand. Functional magnetic resonance imaging (fMRI) studies have shown that the sound input in such cases activates the hearing area in the temporal lobe, but there is also an overlapping phenomenon so that the sensory cortex is also activated [10] (Fig. 10.7). Thus, the brain adapts to the conditions so that the sound from the Sensor Glove is perceived as a true tactile experience.

Synaesthesia

Synaesthesia is the most dramatic form of interaction among senses. Synaesthesia means that when there is input from one sense, other senses become activated. In cases of synaesthesia, a person can literally feel the taste of a shape or a colour [12, 13]. Smells can be perceived as colours, sounds as contours and contours as smells

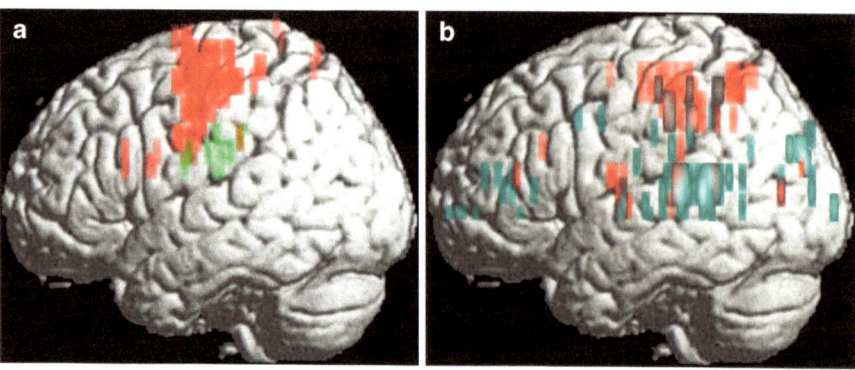

Fig. 10.7 Normally tactile input activates the somatosensory cortex (*red*), while auditory input activates the cortical auditory areas (*green*) (**a**). In this investigation, touch stimuli are applied to the hand of a trained subject. The hand is fitted with a Sensor Glove supplied with miniature microphones. Touch stimuli activates both the hearing area and the somatosensory cortex (**b**) – the auditory input is perceived by the brain as a tactile phenomenon. The pictures are based on an investigation using fMRI technology carried out 2004 by Thomas Hansson and his research team, Linköping University, Sweden (Courtesy of Thomas Hansson)

[14]. Seeing shapes or letters can induce specific colour experiences, and even listening to music or hearing special words can give rise to colour perceptions. When listening to music, various keys can have specific colours and smells. Light can smell and colours can be tasted. It is believed that synaesthesia may occur when connections between the cortical areas for separate senses are especially well developed [15].

The Rubber Hand Phenomenon

Can the sensibility of the hand be transposed into a foreign object? Yes! The so-called rubber hand phenomenon shows how sensibility and vision collaborate in such a way that hand sensibility can be transposed into an artificial hand or another foreign object [16–20]. In a typical experiment, the test subject sits at a table with his right hand hidden behind a screen, out of sight. A rubber hand is placed on the table in front of him in the same position as the right hand would have been in if it were not hidden behind the screen. The examiner is positioned on the other side of the table. With both hands he strokes over the rubber hand and the hidden hand exactly synchronously while the test person, under deep concentration, observes the rubber hand (Fig. 10.8). Within about 30 s a strange phenomenon occurs; to the test person's big surprise, there is a fusion of visual input and tactile sensory input. The sensibility of the test person's hand is transposed to the rubber hand, which the subject suddenly perceives as a part of his own body. If a knife suddenly approaches the rubber hand, it is perceived as a real threat – the subject reacts with increased

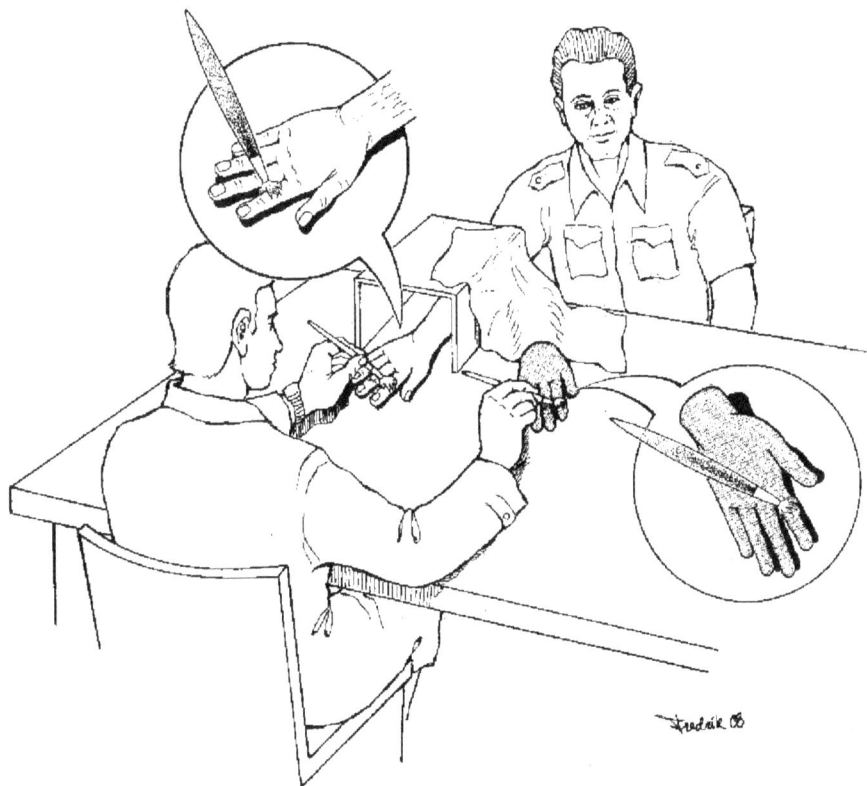

Fig 10.8 The rubber hand phenomenon. The test subject is positioned at a table with his right hand out of sight, hidden behind a screen. A rubber hand is placed in front of him in the same position as his right hand would normally have been. The examiner is positioned on the other side of the table. With both hands, the examiner strokes the index finger of the rubber hand and the index finger of the hidden hand at the same time, while the subject concentrates on observing the rubber hand. Within about 30 s, the sensibility of the subject's hand is transposed to the rubber hand, which he suddenly perceives as a part of his own body

sweating and rapidly withdraws the hand [21, 22]. The principle works even if the rubber hand is replaced with a robotic hand prosthesis [23] and even if the hidden remaining arm stump of an amputee, rather than an intact hand, is stroked synchronously with stroking the robotic hand (see Chap. 16).

The phenomenon illustrates how visual impressions exert a very strong effect on the brain's functioning; the brain is quite simply tricked by the visual input occurring simultaneously with the tactile sensory input. The phenomenon is reminiscent of what we can experience watching a skilful ventriloquist; the dummy seems to talk because its mouth moves, even if the sound comes a bit from the side of the ventriloquist's mouth. A similar phenomenon occurs when we watch a televised concert after switching the sound to stereo speakers placed somewhere else in the room. It does not take long before the brain combines the visual and auditory inputs so that the sound is perceived as intimately linked to what is happening on the screen.

Helen Keller

Like no other, Helen Keller demonstrated how the sensibility of the hand can be a key to communication with the outer world [24–27]. Helen was born on 27 June 1880 in Tuscumbia, Alabama. She was a lively and happy baby who had learned to walk and speak by 19 months of age, when she was suddenly struck by an illness, probably encephalitis. After several days of high fever, she recovered, but her parents found, to their great despair, that their daughter could no longer see and didn't react to their voices. The illness had made her blind and deaf.

Helen now had to explore the environment in her own way. She used the sensibility of her hands to touch objects and to understand their shape and texture, and she learned how to differentiate between smells of various objects. But the previously happy and lovely child had become transformed into a suffering, isolated child enclosed in a dark silence. She became easily aggressive, kicked, bit and pinched people in the vicinity, broke china, refused to be dressed and stole food from other plates on the table.

Helen's salvation was Anne Sullivan, a former student at the Perkins Institute in Boston, where she had learned how to teach deaf-blind children. Anne Sullivan came to Tuscumbia in March 1887 at 21 years of age, when Helen was barely 7. Helen later described this day as 'my soul's birthday'. Before this date she had just existed – now she was beginning to live.

The real breakthrough, breaking Helen's isolation, occurred on 5 April 1887. The key was to utilise the sensibility in the palm of the hand. Helen suddenly understood that everything has a name and that she could learn these names by Anne spelling the words in Helen's palm. On this special day, they both went out to the water pump, and Anne made Helen keep her hand under the pouring water while she pumped. While cold water streamed over one of Helen's hands, Anne spelled 'w-a-t-e-r' in the palm of the other hand. A sudden insight came over Helen and she understood the context – how the words being drawn in her hand corresponded with specific items in the environment.

From this day on, Anne kept spelling words in Helen's palm while Helen touched corresponding objects with her free hand, thereby understanding the meaning and the contents of words. She rapidly learned to write and read the Braille alphabet, and she also learned how to communicate with other people by placing the first two fingers of one hand on the person's lips and her thumb on the person's throat. In this way she could feel the lip movements and the vibrations generated by the spoken words.

Helen and Anne lived an inseparable life, and despite her disability Helen gained a worldwide reputation as an author, feminist and supporter of the disabled. In 1904, she received a Bachelor of Arts degree at Radcliffe College. Helen Keller died 1968 at the age of 88.

References

1. Bavelier D, Neville HJ. Cross-modal plasticity: where and how? Nat Rev Neurosci. 2002;3(6):443–52.
2. Macaluso E, Frith CD, Driver J. Modulation of human visual cortex by crossmodal spatial attention. Science. 2000;289(5482):1206–8.
3. Pascual-Leone A, Hamilton R. The metamodal organization of the brain. Prog Brain Res. 2001;134:427–45.
4. Johansson RS, Westling G, Backstrom A, Flanagan JR. Eye-hand coordination in object manipulation. J Neurosci. 2001;21(17):6917–32.
5. Sadato N, Pascual-Leone A, Grafman J, Ibanez V, Deiber MP, Dold G, et al. Activation of the primary visual cortex by Braille reading in blind subjects. Nature. 1996;380(6574):526–8.
6. Gizewski ER, Gasser T, de Greiff A, Boehm A, Forsting M. Cross-modal plasticity for sensory and motor activation patterns in blind subjects. Neuroimage. 2003;19(3):968–75.
7. Lundborg G, Rosen B, Lindberg S. Hearing as substitution for sensation: a new principle for artificial sensibility. J Hand Surg Am. 1999;24(2):219–24.
8. Rosen B, Lundborg G. Early use of artificial sensibility to improve sensory recovery after repair of the median and ulnar nerve. Scand J Plast Reconstr Surg Hand Surg. 2003;37(1): 54–7.
9. Lundborg G, Rosen B. Enhanced sensory recovery after median nerve repair: effects of early postoperative artificial sensibility using the sensor glove system. J Hand Surg Am. 2003;28 Suppl 1:38–9.
10. Lundborg G, Bjorkman A, Hansson T, Nylander L, Nyman T, Rosen B. Artificial sensibility of the hand based on cortical audiotactile interaction: a study using functional magnetic resonance imaging. Scand J Plast Reconstr Surg Hand Surg. 2005;39(6):370–2.
11. Rosen B, Lundborg G. Enhanced sensory recovery after median nerve repair using cortical audio-tactile interaction. A randomised multicentre study. J Hand Surg Eur Vol. 2007;32(1): 31–7.
12. Ramachandran VS, Blakeslee S. Phantoms in the brain: human nature and the architecture of the mind. London: Fourth Estate; 1999.
13. Ramachandran VS. The tell-tale brain: unlocking the mystery of human nature. London: William Heinemann; 2011.
14. Cytowic RE. The man who tasted shapes. Cambridge: MIT Press cop; 2003.
15. Robertsson D. Learn to see sounds and hear colours. New Scientist. 2009. p. 13.
16. Botvinick M, Cohen J. Rubber hands 'feel' touch that eyes see. Nature. 1998;391(6669):756.
17. Armel KC, Ramachandran VS. Projecting sensations to external objects: evidence from skin conductance response. Proc Biol Sci. 2003;270(1523):1499–506.
18. Ehrsson HH, Spence C, Passingham RE. That's my hand! Activity in premotor cortex reflects feeling of ownership of a limb. Science. 2004;305(5685):875–7.
19. Ehrsson HH, Holmes NP, Passingham RE. Touching a rubber hand: feeling of body ownership is associated with activity in multisensory brain areas. J Neurosci. 2005;25(45):10564–73.
20. Tsakiris M, Haggard P. The rubber hand illusion revisited: visuotactile integration and self-attribution. J Exp Psychol Hum Percept Perform. 2005;31(1):80–91.
21. Ehrsson HH, Wiech K, Weiskopf N, Dolan RJ, Passingham RE. Threatening a rubber hand that you feel is yours elicits a cortical anxiety response. Proc Natl Acad Sci U S A. 2007; 104(23):9828–33.
22. Yong E. Out-of-body experience: master of illusion. Nature. 2011;480(7376):168–70.
23. Rosen B, Ehrsson HH, Antfolk C, Cipriani C, Sebelius F, Lundborg G. Referral of sensation to an advanced humanoid robotic hand prosthesis. Scand J Plast Reconstr Surg Hand Surg. 2009;43(5):260–6.
24. Castor H. Helen Keller. Londons: Franklin Watts Ltd; 1998.
25. Lynch E. The life of Helen Keller. Oxford: Heinemann Library; 2006.
26. Hanks G. Helen: the story of Helen Keller. Exeter: Religious Education Press; 1977.
27. Nielsen KE. The radical lives of Helen Keller. New York: New York University Press; 2004.

Chapter 11
Mirrors in the Brain

Abstract *Mirror neurons* are motor neurons that are activated when we perform a motor action and when we observe a motor action performed by someone else. The first article on mirror neurons, published by Rizzolatti in 1996, inspired new theories about mechanisms for learning by imitation, action understanding and the ability to feel empathy. Mirror neurons are mainly found in the premotor cortex in the frontal and parietal lobes and have strong connections to the cortical visual and emotional areas in the limbic system. Additionally, observation of sensory events can be reflected in our own sensory cortex; observing the touch of a hand activates the hand's representational area in the sensory cortex. These areas can also be activated by listening to 'action words' associated with hand activities and by just imaging a hand activity – so-called motor imagery.

The term 'mirror neurons' refers to types of nerve cells in the brain that are activated when we perform a motor activity or when we observe motor actions performed by others. The discovery of mirror neurons has greatly influenced our view on how the brain functions. About 20 years ago, Giacomo Rizzolatti, professor of physiology at Parma University, accidently discovered the existence of mirror neurons – nerve cells in our brains that can be said to mirror external actions [1]. In laboratory experiments on macaque monkeys, the electrical activity was recorded in the part of brain cortex involved in planning motor actions, the so-called premotor cortex, located in the frontal lobe. The electrical activity in nerve cells in this area could be registered and listened to as crackling sounds in a loudspeaker via electrodes inserted in a specific area called F5. When the monkey stretched out an arm to grasp a peanut, there was a sudden powerful crackling sound from the loudspeaker. When the monkey rested the arm, the sound ceased, but when the monkey again reached for a peanut, the sound reappeared.

Something completely unexpected happened when the experiment was finished and one of the scientists was cleaning up the table. The monkey did not move at all but intensely observed what was happening. When the scientist picked up a peanut, to everyone's surprise, there was a sharp crackling sound from the loudspeaker;

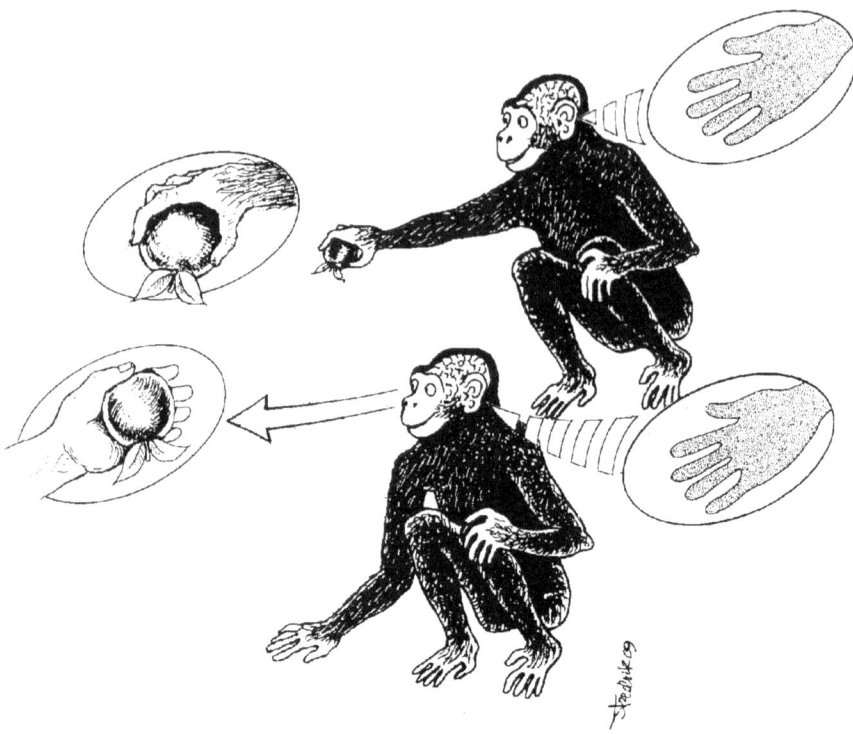

Fig. 11.1 In experiments on macaque monkeys in 1996, the Italian scientist Giacomo Rizzolatti demonstrated that specific nerve cells in the monkey's brain were activated not only when the monkey grasped an object but also when the monkey observed another hand grasping the object. He named these nerve cells 'mirror neurons' – they 'mirror' in the monkey's brain motor activities that are performed by someone else. In the picture, cortical hand representational areas in the premotor cortex are activated when the monkey grasps a fruit as well as when he observes another (Illustration: Fredrik Johansson)

nerve cells in the monkey's brain were responding to the scientist's hand movement (Fig. 11.1). Thus, the effect was the same whether the monkey performed the action or observed someone else doing it. The conclusion was that the brain contained nerve cells that were activated when the monkey performed an active hand movement as well as when it observed a hand movement performed by someone else. These cells were named 'mirror neurons' since they mirrored – in the monkey's brain – motor actions in its surroundings.

Imitation and Learning

In Rizzolatti's laboratory experiment, it was obvious that cortical hand representational areas are activated by the mere observation of others' hands in action. But the occurrences of mirror neurons in the brain are of much greater and more general importance. After the first article was published 1996, the mirror phenomenon in the brain inspired new theories about mechanisms for imitation, learning, action understanding and even sympathy and the ability to feel empathy [1–4]. The monkey experiment showed that just by observing someone picking up a peanut, the monkey automatically generated a copy of the movement in its own head. Today, we know that humans have an even more advanced system of mirror neurons that are distributed within areas of the brain cortex that handle movements, language and feelings. They are nerve cell populations in the brain mainly localised in the premotor cortex in the inferior parts of the frontal and parietal lobes, with strong connections to the visual area in the occipital lobe and emotional areas in the limbic system [5].

Man's mirror neurons are important for learning by imitation; they are of central importance in learning complex motor tasks like tying shoelaces or a tie or learning the chord grips on a guitar by observing the teacher [6]. Basically, the mirror neurons create a representation of other individuals' actions in the observer's own motor system, just as if the observer were performing an identical action himself [7, 8].

Thus, mirror neuron incidents occurring in our surroundings are mirrored and 'imitated' in our minds. When we watch a high-jump competition, enjoying the Olympics on TV, most of us feel obvious muscle contractions in our right leg when the high jumper takes off towards the bar. Several experiments have also demonstrated that our mirror neurons help us to share other individuals' experiences and feelings as a result of watching their facial expressions. Consequently, a biological base for empathy is generated, and thus one may explain the infectiousness of yawning, laughter and also many types of positive or negative moods. Few of us can resist yawning when we observe another person yawning.

Mirror neurons can explain much about how we learn to walk, act, dance or play tennis. But on a deeper level, they also play a role in our understanding of other people's feelings and problems. When we observe people in our vicinity or even on TV, in movies or the theatre experiencing pain, it activates our own pain centre: observing other people's reactions like discomfort or disgust can activate corresponding centres in our own brains [9–12]. Perhaps this explains why some people more easily feel empathy than others do. Autistic children have an impaired capacity for activating their mirror neurons, and it has been proposed that they therefore have difficulties in communicating with others and difficulties in understanding others' intentions and the meaning of their motor activities and facial expressions – 'broken mirror syndrome'. Psychopaths, too, seem to be dysfunctional in this context, with a weakened activity pattern among the mirror neurons.

Understanding the Intentions of Others

The mirror neurons also make it possible to understand the intentions behind other people's body language and movements: the mirror function and the activation of familiar, well-established motor patterns in the observer's brain automatically generate an understanding of the meaning and intentions of movements performed by others – *action understanding* [7, 8]. Some groups of mirror neurons are activated by observing simple movements in the hand – stretching out the arm, grasping an apple or picking up a peanut – while other groups of mirror neurons are more selective in their reaction pattern and become activated in response to the presumed meaning [13, 14]. In an experiment by Iacoboni, test subjects were asked to observe pictures of people reaching for various items on a tea tray – a teapot, a mug, a cream pitcher, a basket of biscuits and some napkins. Various groups of mirror neurons were activated depending on the intention of the movement; if the hand was presumed to lift a cup of hot tea for drinking purposes, it triggered one group of mirror neurons, but quite another population was activated if the observed hand was supposed to lift the cup in order to clean up and wipe the table [13]. Thus, the activation of mirror neurons requires an understanding of the meaning and intention of the observed movement. When a monkey clearly understands the meaning of an observed hand function, the mirror neurons are activated even if the movement is not completed, for instance, if the last part of the activity is hidden behind a screen [14, 15].

Thus, we can intuitively understand other people's feelings, actions and intentions, thanks to mirror neurons [16]. In team sports, the group synergy and intuitive understanding of one another's movements on the field are important to the success of the team. Here we can see much of the value in active tutoring and practising in groups. In rehabilitation programmes, especially after hand injuries, it is an advantage to plan the training on a group basis, making it possible to activate the motor system through the patient's own activities and also by observing other members' motor actions.

Mirror neurons are active early in a child's development: if a mother shows the tip of her tongue to her newborn baby, she immediately gets a mirror response. They are also at work in communication between humans and animals. In Rizzolatti's original experiments, a monkey's mirror neurons were activated by the human's hand movements. It does not matter if the hand is close to or at some distance from the monkey – apparently the hand's size is not important.

When you cuddle with a kitten, the animal will extend its tongue if you stick out your own. While I was on a safari tour in Tanzania, the jeep drove up close to a big male lion at just a few metres' distance. One of the participants, Gabriela, locked eyes with the big lion. Gabriela blinked at the lion and immediately got a blink in return – a breathtaking communication between two individuals who, for a moment, were mirrored in each other's brains.

Mirroring Touch

Mirror nerve cells are activated by the observation of motor actions and by the observation of sensory experiences: the mere observation of someone else's hand or leg being touched stimulates the observer's hand or leg area in the sensory cortex [17, 18]. Watching a big poisonous spider advancing on James Bond's arm in a thrilling film activates the same sensory areas in the observer's brain as if the spider was actually crawling up the observer's own arm. Sometimes the sensory cortex can be activated by just observing a characteristic, familiar surface or texture; a sculptor once told me that

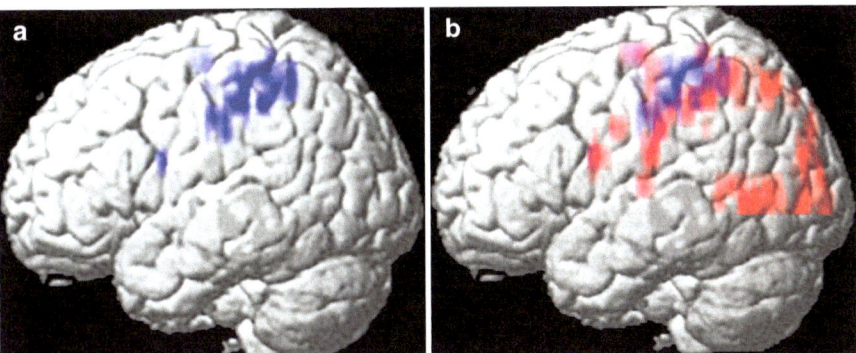

Fig. 11.2 (**a**) Tactile stimulation of the hand activates the somatosensory cortex (*blue*). (**b**) Observing a hand being touched activates the cortical visual areas (*red*) as well as the somatosensory cortex. In this experiment, using fMRI, the subject watches a film sequence where another person's hand is stroked with a brush. The cortical visual area in the occipital lobe is activated, and there is also activation of the somatosensory cortex – observation of touch is perceived as 'real' touch. The experiment is performed 2008 by Thomas Hansson and his research team, Linköping University, Sweden (Courtesy of Thomas Hansson)

when he observed sculptures of various materials, he felt the character of the material and the surface in his mind as though he was actively touching the sculpture.

The effects of mirror neurons in the sensory cortex can be demonstrated in fMRI studies: observing another person's hand being touched naturally activates the visual cortical areas, as well as the hand representational area in the sensory cortex (Fig. 11.2) [18]. This phenomenon can be useful in the rehabilitation and training of the hand after a nerve injury in the arm. When the hand lacks sensibility for some time, its representation in the brain is 'erased'. Watching hands being touched can probably contribute to maintaining the hand's normal location in the sensory cortex until new nerve fibres reinnervate the hand and sensibility returns.

Action Words

Thus, the hand's representational area in the brain is activated not only when one performs a hand movement but also when one observes a hand movement performed by someone else. In fact, it may be enough to read a text containing words that are linked to hand movements – 'action words' – to activate the hand area. fMRI studies conducted by Hauk et al. showed that reading 'action words' that are associated with a hand activity activates the same area in the brain that is activated when the activity is performed (Fig. 11.3) [19–21]. The same phenomenon occurs when listening to someone reading action words aloud. Naturally the auditory area is the first to be activated, but after a few hundredths of a second, the hand area in motor cortex is also activated.

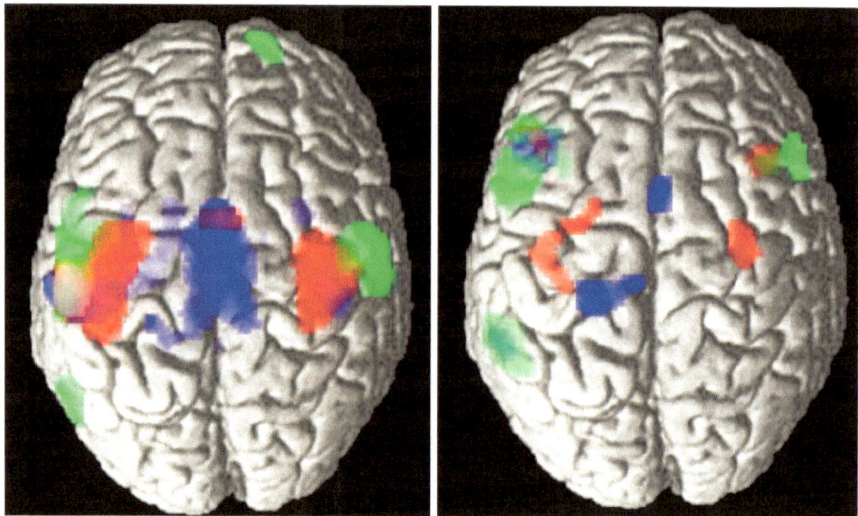

Fig. 11.3 Activation of the hand representational area in the motor cortex (*red*) by reading 'action words' describing hand activities. The same area is activated when a person reads about hand activities (*right picture*) as when the activities are actually being performed (*left picture*). The pictures are based on a study using fMRI (From Hauk et al. [19])

Action Sounds

Observing a hand movement or a hand being touched activates areas in the brain cortex that are also activated when we perform the hand movement or when our own hand is touched. The mere sound of a hand or mouth movement activates cortical hand areas that are activated when we perform the corresponding movement. This phenomenon is based on *auditory mirror neurons* that are activated by active movements in the hand or mouth and by the sound of such actions [22]. An example of such an action sound is the rustling when we crumple a paper bag. Perhaps we have here the secret behind some people's special ability to imitate – mirroring another person's speech in the impersonator's auditory mirror neurons?

Motor Imagery

When preparing for an athletic performance, it may be useful to 'imagine' the whole procedure and mentally go through all phases of the activities. Likewise, musicians can mentally 'perform' a music piece and thus prepare hand and brain for the task by going through the complete motor programme [23].

There is a physiological basis for such 'motor imagery' of a motor task. fMRI studies have demonstrated that an imagined movement of a finger, a hand or a foot activates the same area in the brain cortex as if the movement is in fact being

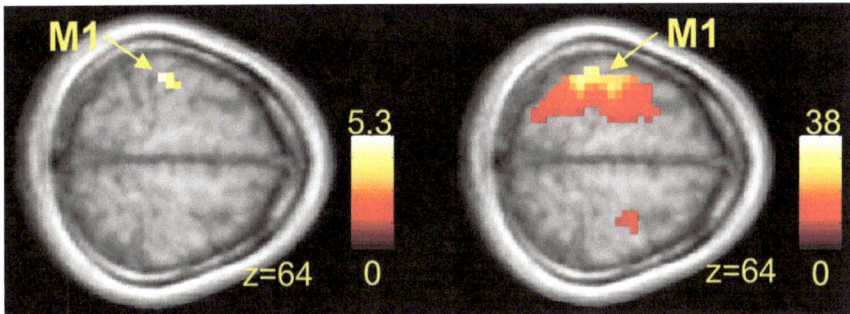

Fig. 11.4 'Motor imagery'. Imagining a hand movement (*left picture*) activates the same area in the motor cortex (*M1*) as if the movement is actually performed (*right picture*). The picture is based on a study using the fMRI (From Ehrsson et al. [24])

performed (Fig. 11.4) [24]. Again, this might possibly help the brain to maintain the hand's representational area when the hand has not received any sensory stimuli for an extended time period, such as during the first phase after repair of an injured nerve in the forearm before the hand recovers its sensory functions.

Hand and Mouth in Cooperation

Mirror neurons are richly present in Broca's area, the area in the brain's frontal lobe that is essential for speech. Broca's area has overlapping areas associated with the mouth and the hand, giving them a close natural connection in our brains (Fig. 11.5) [25]. Hand movements, performed under deep concentration, for example, when writing by hand, are often accompanied by synchronous movements of the tongue and lips. Perhaps singing and chatting during work facilitate the activities and movements of the hands. Among seasoned actors, body language and rehearsed gestures may help in remembering their lines during a performance – perhaps an illustration of a link and a common substrate for movements of the tongue and hand.

The close link between hand and mouth movements was demonstrated in several neurophysiological studies; hand/arm movements and speech are closely associated in the brain [25]. The excitability in the hand's representation in the motor cortex increases in association with speech and reading. In clinical studies, test subjects were instructed to bite down on a small or big object and at the same time open their right hand. Spontaneously the subjects opened their hand widest when biting down on the big object [26, 27]. When a syllable is spoken while the hand simultaneously grips an item, gripping a big item may make the mouth open wider than if the hand grips a smaller item. The same thing may happen to the mouth of someone watching someone else picking up large or small items – an effect presumably associated with the mirror neurons in Broca's area.

Fig. 11.5 Hand and brain in cooperation. A balloon salesman makes a knot to tie the balloons together, making good use of his mouth (Photo: Jan Delden)

References

1. Rizzolatti G, Fadiga L, Gallese V, Fogassi L. Premotor cortex and the recognition of motor actions. Brain Res Cogn Brain Res. 1996;3(2):131–41.
2. Iacoboni M, Woods RP, Brass M, Bekkering H, Mazziotta JC, Rizzolatti G. Cortical mechanisms of human imitation. Science. 1999;286(5449):2526–8.
3. Gallese V, Fadiga L, Fogassi L, Rizzolatti G. Action recognition in the premotor cortex. Brain. 1996;119(Pt 2):593–609.
4. di Pellegrino G, Fadiga L, Fogassi L, Gallese V, Rizzolatti G. Understanding motor events: a neurophysiological study. Exp Brain Res. 1992;91(1):176–80.
5. Mukamel R, Ekstrom AD, Kaplan J, Iacoboni M, Fried I. Single-neuron responses in humans during execution and observation of actions. Curr Biol. 2010;20(8):750–6.
6. Buccino G, Vogt S, Ritzl A, Fink GR, Zilles K, Freund HJ, et al. Neural circuits underlying imitation learning of hand actions: an event-related fMRI study. Neuron. 2004;42(2):323–34.
7. Bekkering H. Imitation: common mechanisms in the observation and execution of finger and mouth movements. In: Meltzoff AN, Prinz W, editors. The imitative mind: development, evolution and brain bases. Cambridge: Cambridge University Press; 2002.

8. Byrne RW. The thinking ape: evolutionary origins of intelligence. Oxford: Oxford University Press; 1995.
9. Keysers C, Gazzola V. Expanding the mirror: vicarious activity for actions, emotions, and sensations. Curr Opin Neurobiol. 2009;19(6):666–71.
10. Keysers C, Kaas JH, Gazzola V. Somatosensation in social perception. Nat Rev Neurosci. 2010;11(6):417–28.
11. Bastiaansen JA, Thioux M, Keysers C. Evidence for mirror systems in emotions. Philos Trans R Soc Lond B Biol Sci. 2009;364(1528):2391–404.
12. Ramachandran VS. The tell-tale brain: unlocking the mystery of human nature. London: William Heinemann; 2011.
13. Iacoboni M. Neural mechanisms of imitation. Curr Opin Neurobiol. 2005;15(6):632–7.
14. Fogassi L, Ferrari PF, Gesierich B, Rozzi S, Chersi F, Rizzolatti G. Parietal lobe: from action organization to intention understanding. Science. 2005;308(5722):662–7.
15. Fogassi L, Gallese V, Buccino G, Craighero L, Fadiga L, Rizzolatti G. Cortical mechanism for the visual guidance of hand grasping movements in the monkey: A reversible inactivation study. Brain. 2001;124(Pt 3):571–86.
16. Iacoboni M. Mirroring people: the new science of how we connect with others. New York: Farrar, Straus and Giroux; 2008.
17. Keysers C, Wicker B, Gazzola V, Anton JL, Fogassi L, Gallese V. A touching sight: SII/PV activation during the observation and experience of touch. Neuron. 2004;42(2):335–46.
18. Hansson T, Nyman T, Bjorkman A, Lundberg P, Nylander L, Rosen B, et al. Sights of touching activates the somatosensory cortex in humans. Scand J Plast Reconstr Surg Hand Surg. 2009;43(5):267–9.
19. Hauk O, Johnsrude I, Pulvermuller F. Somatotopic representation of action words in human motor and premotor cortex. Neuron. 2004;41(2):301–7.
20. Pulvermuller F, Hauk O, Nikulin VV, Ilmoniemi RJ. Functional links between motor and language systems. Eur J Neurosci. 2005;21(3):793–7.
21. Hauk O, Shtyrov Y, Pulvermuller F. The time course of action and action-word comprehension in the human brain as revealed by neurophysiology. J Physiol Paris. 2008;102(1–3):50–8.
22. Pazzaglia M, Pizzamiglio L, Pes E, Aglioti SM. The sound of actions in apraxia. Curr Biol. 2008;18(22):1766–72.
23. Jeannerod M. The representing brain: neural correlates of motor intention and imagery. Behav Brain Sci. 1994;17:187–245.
24. Ehrsson HH, Geyer S, Naito E. Imagery of voluntary movement of fingers, toes, and tongue activates corresponding body-part-specific motor representations. J Neurophysiol. 2003;90(5):3304–16.
25. Rizzolatti G, Sinigaglia C. Mirrors in the brain: how our minds share actions and emotions. Oxford: Oxford University Press; 2008.
26. Meister IG, Boroojerdi B, Foltys H, Sparing R, Huber W, Topper R. Motor cortex hand area and speech: implications for the development of language. Neuropsychologia. 2003;41(4):401–6.
27. Gentilucci M, Benuzzi F, Gangitano M, Grimaldi S. Grasp with hand and mouth: a kinematic study on healthy subjects. J Neurophysiol. 2001;86(4):1685–99.

Chapter 12
Creative Hands

Abstract Creative hands can be traced back to cave art, decorations and various types of artistically designed objects created over the last 100,000 years, primarily in South Africa and Europe. As a symbol of creativity, the hand plays a key role in our culture, particularly in art and music. However, many of the most well-known artists and musicians suffered from various types of hand problems but were still able to perform their art, thanks to well-preserved brain function – their creative capacity. Musicians lacking one arm have been able to perform music at very high artistic levels. The brains of experienced musicians may be somewhat different from the brains of nonmusicians, with well-developed white matter and an enlarged corpus callosum connecting the two brain hemispheres, and in violin players the cortical representation of the left hand may be enlarged. Experienced musicians may view their instrument as an integral part of their bodies.

When did the ability for creative thinking and the desire to decorate and beautify items and tools of everyday life emerge among our ancestors? And when did the desire arise to shape and design items for self-decoration? Creative hands can be traced back to the development of man. Some of the oldest known artistically designed items are a couple of carefully polished pieces of ochre with engraved geometric patterns characterised by regular engraved crossed lines that were found in the Blombos cave in South Africa. It has been proposed that these nearly 77,000-year-old items bear witness to artistic thinking and an early symbolic tradition [1]. Thirteen 100,000-year-old engraved ochre pieces were recently discovered in the same cave together with a processing workshop where a liquefied ochre-rich mixture was produced and stored in abalone shells [2]. In the Blombos cave, scientists also discovered the oldest known jewellery made of shells with drilled holes to wear as a necklace [3]. The findings from the Blombos cave indicate that a symbolic tradition existed among our ancestors 100,000 years ago or even earlier. Paul Mellars at the University of Cambridge argues that the ability for aesthetic and symbolic thinking might have existed since the emergence of our species *Homo sapiens* 170,000–200,000 years ago. But no one can understand and interpret the meaning and symbolism of these items for certain.

G. Lundborg, *The Hand and the Brain*,
DOI 10.1007/978-1-4471-5334-4_12, © Springer-Verlag London 2014

An interesting example of the use of a type of very early graphic art for communication purposes is symmetric engravings on 60,000-year-old ostrich eggs. For 10 years, Pierre-Jean Texier at Bordeaux University and his associates collected 270 such ostrich eggs at the Diepkloof Rock Shelter, a cave in west South Africa. Various engravings were repeated several times on the egg shells over long time periods. Scientists interpreted this as an indication of symbolic communication among individuals, a 'modern' human behaviour. The eggs were most likely used as containers, and the signs might have symbolised either the content of the container or the name of its owner [4].

It was long believed that the most evident proof of high-level aesthetic thinking by our ancestors is the pictures and paintings that were discovered in the caves of southern Europe and at many other locations worldwide. Compared to the findings from the Blombos cave, the European cave paintings are considerably younger – they have been dated at 10,000 to 40,000 years old. These paintings were made using red and yellow ochre, hematite, magnesium oxide and charcoal. Among the oldest paintings are those in the Chauvet cave in southern France, made 32,000 years ago. This cave contains a rich gallery of animal images. In some locations the silhouettes of animals were etched in the cliff wall before they were coloured. The most common motifs are large wild animals like aurochses, bison, lions, bears and rhinoceroses, but there also are pictures of wild horses, reindeer and stags [5]. In the Chauvet cave there are spectacular images of lions that were painted in harmony with protruding irregularities from the wall so that the animals almost seem to emerge from the cliff wall [6]. Images of animals also dominate the cave at Niaux in the French Pyrenees [7].

The animal images can be interpreted as a type of hunting magic. The images most often depict the animals that they hunted, but sometimes they represent animals that competed with the hunter and predators that were very dangerous to the hunters. The animal pictures are not static icons but represent animals running, swimming, hunting, fighting, copulating and giving birth.

Prominent among several other caves with paintings showing a high level of artistry is the Lascaux cave in Dordogne in southern France (Fig. 12.1). It was accidentally discovered by some 13-year-old boys in September 1940. This cave, containing thousands of old paintings and engravings about 17,000 years old, has been called 'the Sistine Chapel of prehistory' and is on UNESCO's World Heritage List. But the images have been severely damaged, possibly due to climate changes or all the carbon dioxide exhaled by many visitors over the years. Massive attacks of algae caused the paintings to suffer from fungus that spread dark patches over the cave walls. For many years intensive efforts have been made to reverse this development, and 30 years ago the cave was closed in order to return it to its original condition [8].

The hand is a symbol of creativity and spiritual power. For many people the most creative and powerful hand is the hand of God, which in the religious perspective created and modelled the earth, heaven and man. On the ceiling of the Sistine Chapel, God's finger, as painted by Michelangelo, symbolises the basis for the creation of man – God's right index finger transfers the spark of life to Adam by touching his left hand (Fig. 12.2).

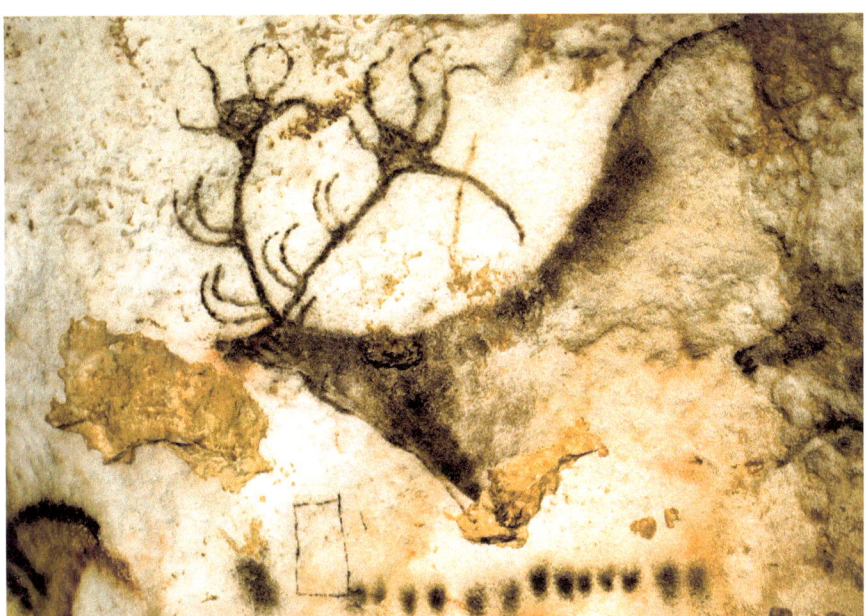

Fig. 12.1 Works by early creative hands. A cave painting from the Lascaux cave, about 20,000 years old, representing a bellowing deer with enormous antlers. The meaning of the *dots* and *rectangle* below the painting is not known (From *National Geographic* Oct 1988. Photo: Sisse Brimberg/ National Geographics)

Fig. 12.2 God transmits the spark of life to Adam. Michelangelo, painting on the ceiling of the Sistine Chapel

Fig. 12.3 *The Cathedral,*
Auguste Rodin 1908. Two
hands in prayer form the
peaked arch. No 1001, stone,
64 × 29.5 × 31.8 cm (Photo:
Christian Baraja. Musée
Rodin, Paris)

Hands often have a prominent and important role in paintings and sculptures by the French sculptor Auguste Rodin (Fig. 12.3), and in many of Swedish sculptor Carl Milles' works the shape and posture of the hands contribute to the figures' vitality and expressions.

Carl Milles visited Rodin's studio in Paris in 1902 and was deeply touched by a marble sculpture representing God's hand creating man (Fig. 12.4). Milles was inspired to create a sculpture on the same theme, however, symbolising man's existence in God's hand with man balancing on the thumb and index finger (Fig. 12.5).

Every day we see creative hands all around us. It is easy to be impressed by the creative power and skill of hands in many activities and occupations where handwork has not yet been replaced by automatic and computerised processes. The silent knowledge, possibilities and improvisation capacity of hands are still the basis for several occupations, even if many handicraft traditions and inherited knowledge are at risk of disappearing. We can still be aware of the impressive handicraft skills exhibited by carpenters, blacksmiths, shipbuilders and metalworkers as well as bakers, hairdressers, instrument makers, cooks and florists. The hand plays a very special role in handicrafts, art and music, constituting a prerequisite for the artistic creation processes. The hand's relation to light and shadow create special possibilities (Fig. 12.6).

Fig. 12.4 *The Hand of God.*
Auguste Rodin 1908. No 988,
marble 95.5×75×56.5 cm.
God's hand modelling man
(Photo: Christian Baraja.
Musée Rodin, Paris)

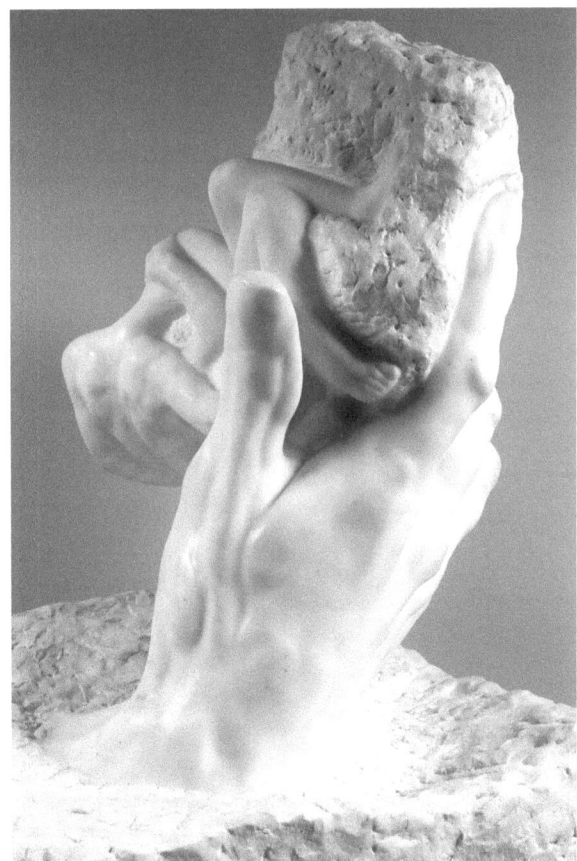

Does creativity reside in the hand or in the brain? Musicians may feel that their hand has learned how to create the music, that it 'knows what to do' and that the creative process is actually located in the fingers. One of my patients, a piano teacher, expressed it like this: 'Doctor, it feels as if my brain is located in my fingers.' She had been wondering how she could sometimes be thinking of something completely different while her hand, by itself, was playing etudes and waltzes by Chopin.

The Hand, the Brain and the Creative Mind

Creativity is located in the brain, and the hand can 'play by itself' because experience and training have shaped motor programmes that remember and control the hand's movements. We know that an artist's creativity can still be there even if the hand's function is severely impaired. One example is Pierre-Auguste Renoir who suffered from a rheumatic disease that severely impeded his gripping function and

Fig. 12.5 Man balancing on
God's hand. *The Hand of
God*, 1949–1953, by Carl
Milles (Courtesy of
Millesgården, Stockholm.
Photo: Lars Ekdahl ©
Millesgården, Stockholm)

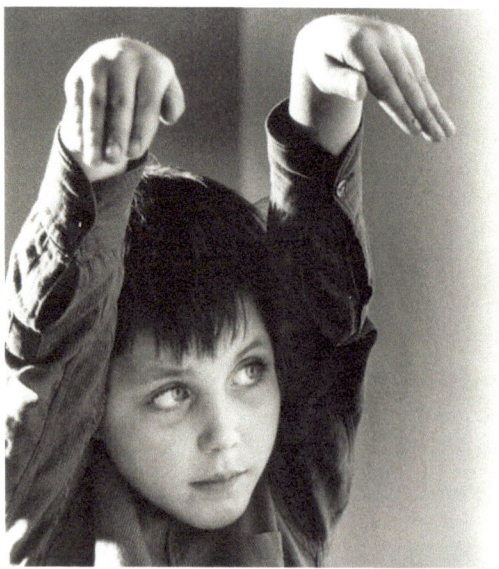

Fig. 12.6 Creative hands of a child making shadow images (From Claude Verdan, *La main, cet univers*. Fondation du Musée de la Main, Lausanne, 1994. © Monique Jacot)

hand precision. Renoir had to tie the paintbrush to his hand to be able to paint, yet even during periods of severe illness he produced several of his most magnificent paintings. The hand overcame its difficulties despite its very bad condition. Henri Matisse was very impressed by this when he wrote the following about Renoir:

> A long lasting suffering – his finger joints were all swollen and severely malformed. – and still he painted his best masterpieces! While his body was fading away his soul seemed to grow stronger and he expressed himself more and more easily [9]

The painter Raoul Dufy was also afflicted with rheumatoid arthritis, but in his case it negatively affected his painting. His capacity to draw and paint gradually became worse as his hand stiffened [9]. But, interestingly enough, medical advances caught up with the disease – effective treatments were developed, and he could again create his art as a direct effect of improved hand function. At age 73, Dufy was severely limited by his illness, but he was one of the first patients to test treatment with cortisone. The result was dramatic; after only a few days of treatment, the mobility and power of his fingers improved. For the first time in several years, he could squeeze out paint from the tubes without help. Once again he could master his tools and create high-quality paintings.

Without treatment, gradual impairment of hand function can negatively affect art. Paul Klee was one of our time's most talented and special artists who enjoyed constantly creating new pictures in new colours and new constellations. At the age of 40, Klee was affected by a serious connective tissue disease, *scleroderma*, leading to contractures in the hand and stiffness of the finger joints. As a result his art lost its joyful character, and there was often a shift towards simple pencil drawings [9].

However, there are examples of how a disease affecting the hands can be an advantage and facilitate artistic expression. The legendary violinist Paganini was technically brilliant and was unsurpassed in playing extremely difficult solo pieces and passages. His way of playing had an almost magical power that had an immense impact on his audiences. It was said that Paganini was closely associated with the devil, and for this reason he was not allowed to be buried on holy ground.

Paganini suffered from a very uncommon connective tissue disease, *Ehler-Danlos syndrome*, characterised by extreme elasticity and increased mobility of the wrist and finger joints. It has been suggested that this contributed to his exceptional technical skills, making it possible for him to play extremely advanced double grips, which astonished listeners and made him famous. The hand's range was greatly increased. With his left thumb on the middle of the violin's neck, he could play in the first three positions without changing his grip. Thus, he was able to play very rapid passages with exceptional precision. Paganini played fervently, violently drawing the bow across the strings and playing up and down the scale with admirable speed, the cadenzas flying from his fingers.

As a hand surgeon I have seen many examples of how hand problems critically influence hand creativity. Most cases concerned music students or professional musicians playing both classical music and jazz. Most often their problems were ergonomic and could be resolved by changing body posture and playing technique. However, sometimes it was a disease where anatomical anomalies in the hand made it impossible to reach all of the instrument's valves and controls. Often the problem can be solved by a slight modification of the instrument's construction, compensating for limitations in hand function, but many times this is impossible. Dupuytren's contracture is a pathological condition in the hand characterised by contractures of one or several fingers that cannot be straightened. Most often the contracture appears in the little and ring fingers. One of my patients, a professional flutist, had severe symptoms from his Dupuytren's disease. He was the principal in a symphony orchestra and had great difficulty managing his work since his little finger didn't reach out to the most distant key on the flute. His hand function was fairly good, thanks to several surgical procedures over several years, but eventually his hand could not tolerate any more surgery – luckily this coincided with his retirement. Another patient, a young woman, was taking classes to become an alto-violin player but unfortunately developed a moderate form of psoriasis that affected her distal finger joints. She became a physiotherapist, thus becoming doubly skilled as an alto-violin player *and* physiotherapist.

A hand problem of special interest that I have encountered, especially among flute players, is poor precision and mobility of the right little finger. These patients were unable to individually flex the middle joint of the little finger without simultaneously also flexing the distal joint of the finger. This inability to individually move the little finger's joints caused major problems in their playing and sometimes forced them to discontinue their ongoing training or switch to another instrument. Intensive exercises were not helpful; the pupil was not able to 'find the little finger'. The condition resulted in anxiety and self-reproach since the pupil believed that he was not training properly. The phenomenon, sometimes called 'lazy little finger', is well known among hand surgeons. It is a matter of a slight anatomical anomaly in the little finger with a lack of one of the two flexor tendons that are present in most of us. In the little finger there are usually two flexor tendons, one 'deep flexor tendon', which gives active flexion to the distal joint, and one 'superficial flexor tendon', which gives active flexion to the middle joint. Sometimes, as with the 'lazy little finger', the superficial flexor tendon is absent, making it impossible to flex the middle joint without also flexing the distal joint. This makes handling the flute's complicated system of keys difficult, sometimes making a career as a solo player impossible. Many with this problem have been very relieved to learn that it is not a question of poor training but an anatomical anomaly we can do nothing about.

Do you lack the superficial tendon in your little finger? Try this simple test. Use your left hand to extend all your fingers, except your little finger, into a straightened position – this blocks the function of *all* deep flexor tendons, including that of the little finger, since they all originate from the same muscle. Now try to flex the middle joint of the little finger; if you can, it means that you have a functioning superficial flexor tendon in your little finger and that you – at least from an anatomical point of view – have what it takes to embark on a solo career as a flute player.

One Hand Missing

One of the most dramatic injuries that can affect a musician is the loss of a hand. Yet several famous musicians have suffered such a loss, especially during times of war when a severely wounded extremity sometimes had to be amputated [9]. During the

Napoleonic Wars and the First World War, several well-known pianists lost one arm. But several of them refused to give up their careers, and there was an awakening of interest among composers and pianists for piano music that could be performed with one hand only – usually the left hand [10–12].

Composers like Liszt, Ravel, Scriabin and Prokofiev composed piano music that was to be performed by the left hand – music that was so complex that, for a listener, it could be perceived as being played with both hands. A one-armed pianist certainly needs an enormous purposefulness, a strong will and technical skill to play it. One of the best known examples is the Hungarian pianist Géza Zichy, born in 1849. At the age of 15, Zichy injured his right arm in a hunting accident. It was so bad that his arm had to be amputated at shoulder level. But Zichy, who had already begun a career as a pianist, was firmly determined to fulfil his plans. He wrote dramatic letters to his piano teacher that were to be opened 1 year later: 'In exactly one year from now, if I am not able to play with one hand what other musicians can do with two hands, shoot a bullet through my head.' Zichy was successful in fulfilling his intention – he became a brilliant pianist. One of Vienna's most feared music critics, Eduard Hanslick, wrote: 'The most astonishing thing we have heard in years when it comes to piano playing was presented by a one-armed man – Count Géza Zichy' [11].

Another well-known example is Paul Wittgenstein, born in Vienna in 1887. Wittgenstein had begun his rapidly ascending career as piano soloist with the Vienna Philharmonic when he, being a soldier in the First World War, was shot in his right arm by a sniper. In order to save his life, the arm was amputated. Wittgenstein then spent 2 years as a prisoner of war in Siberia. After returning to Vienna he again started his soloist career, this time as a left-armed solo pianist. Wittgenstein was good friends with Ravel, who wrote several pieces especially for him, including *Piano Concerto Number 2 in D major for the Left Hand*. Today several of the pieces that Wittgenstein performed with his left hand are performed as concert pieces for solo pianists with both hands intact.

Music for the left hand can also mean a lot to pianists with both hands intact but who have lost function in the right hand because of an illness, for instance, a stroke. The 2011 Swedish Nobel Prize Laureate in literature, who was also an excellent pianist, was affected several years earlier by a left-sided stroke that affected his speech and the function in his right hand. Piano pieces written for the left hand became an important part of his rehabilitation process and an essential part of his continuing life as talented pianist.

If solo pieces for piano can be performed with one arm or with both hands, what could be expected from piano pieces written for four hands? When I recently visited the small village of Anacapri on the Mediterranean island of Capri, I visited a small church where for the whole evening two brothers, both of them professional pianists, performed only music written for four hands. The audience enjoyed compositions by Brahms, Liszt and Respighi. It was an electrifying experience; a massive flow of music filled every corner of the little church. It was performed by two pianists who knew each other inside and out, and during the evening they increasingly inspired and egged each other on. When it was time for the final chord in Liszt's second Hungarian rhapsody (*Seconda rapsodia ungherese*), the keyboard was attacked with such force that the brother on the left lost his balance and was very close to falling on the floor. A dumbfounded audience applauded the brothers so intensely that they played many encores.

Brain, Hand and Art

Thus, it seems that changes in the appearance and function of the hand may have both negative and positive influences on creative ability. So, how is the hand's performance and creativity influenced by pathological processes in the brain?

Since creativity is localised to the brain, a brain injury can severely impact art performance, even if the hand has not been affected by disease. An interesting example is the Swedish artist Carl Fredrik Reuterswärd who, at the peak of his career, became paralysed in the right half of his body because of a left-sided stroke. He was not able to use his right hand anymore, but with training and an enormously strong will, he learned how to paint with his left hand. He began creating left-handed drawings and paintings that showed a different character from his earlier art pieces; they now had a more playful and softer appearance. The left hand was now controlled by his right brain hemisphere, which gained a more dominant role, exhibiting a friendlier and more emotional character as compared to the left brain hemisphere. Reuterswärd's personality also changed in a direction that was reflected in his drawings: 'The sharp sarcasm and spiritual performance was replaced by a playful, kind-hearted simplicity' [9].

Mouth and Hand

In instrumental music the hand is the great master that extracts melodies, rhythms and sounds from the instrument either on its own or in interaction with lips, tongue and lungs, depending on the instrument. Thus the music uses a great deal of the brain's resources; the hands, mouth and tongue are represented in the major motor and sensory parts of the brain's body map. A pianist uses motor and sensory hand areas in both brain hemispheres; a wind instrument player also uses the cortical representations of the oral cavity, the lips and the tongue. In Broca's area in the inferior part of the frontal lobe, motor representations for hand and mouth movements overlap and are well integrated. Cooperation of the hand and mouth in music, therefore, is quite natural, like the natural gestures and movements of our hands when we speak and communicate with others. In music there are several examples of how the hand, mouth and tongue unconsciously cooperate: several of the most famous jazz piano players are easily recognised because of their grunting, humming and production of strange sounds that accompany their playing, for instance, Oscar Peterson, Erroll Garner and Keith Jarrett.

A Difficult Task for the Brain

From a neurophysiological perspective, performing a piece of music is an extremely complicated process that we, the listeners, luckily do not have to worry about. But just think of a pianist who performs a piano piece from a sheet of written music, especially if it is done for the first time, *a prima vista* (Fig. 12.7). The musician must

Fig. 12.7 Performing a piano piece is a complex task for the brain. Pressing down the keys requires an advanced fine motor control of the fingers, which is dependent on sensory feedback via the hearing as well as the sense of touch. The picture below to the right illustrates the skin of the finger pulp with various sensory receptors as they are described in Fig. 8.1 (Illustration: Fredrik Johansson)

have previously learned how to understand and interpret the language of notes and then must programme the brain to understand the notes and all other signs on the written sheet. When the piece is performed, the note pictures have to be projected on the retina of the eye and then be transported to the visual area in the brain cortex to be processed and interpreted in a learned programme. The information must then be transferred to the brain's motor cortex, and signals must be sent down the efferent pathways to the forearm and hand muscles that need to press down the keys in correct sequences and with appropriate force. But in order to use a correct, appropriate force when pressing the keys, sensory feedback is needed from the finger pulps to modulate hand function. Auditory input also plays an important role in regulating the force in finger action in this complicated process. In addition, proprioceptive receptors in tendons, muscles and skin must continuously keep the brain informed about the position of the arm, hand and fingers.

All of this is just the beginning: on top of the basic skill, there is the pianist's personal interpretation and expression based on complicated processes in higher brain centres. But, being listeners, we can just sit back and listen, enjoy and be carried away by the music.

Musicians' Brains

So, if creativity is localised to the brain, does a musician's brain look different from the brain of a nonmusician and does it function in a special way? It may be possible to identify some specific differences regarding structure as well as function [13]. In 1995, Gottfried Schlaug and colleagues at Harvard University published an article showing that the *corpus callosum*, which connects the two brain hemispheres, is more substantial, larger and better expressed in musicians, conveying a need for especially good communication between the two hemispheres [14]. In musicians with absolute pitch, there is an enlargement of the *planum temporale* (a part of the auditory cortex), and there is a greater volume of grey matter (amount of brain cells) in the hearing area [15, 16]. Fredrik Ullén, an active concert pianist and brain scientist at the Karolinska Institute in Stockholm, has used functional magnetic resonance imaging technology (fMRI) to compare the brains of professional pianists with other people. He found an obvious relationship between the start of early training and the appearance of the white matter in the brain cortex. In pianists who had been training since early childhood, the white matter, representing the brain's system of pathways, is more extensively developed compared to non-piano players [17].

Changes in the Cortical Hand Representation

Long-term training induces physical changes in the brain. We know that all parts of the body are represented in cortical body maps in both the motor and sensory brain cortices, but this map is experience dependent, dynamic and constantly changing in response to requirements put on the brain. In practised musicians, the sensory and motor cortices occupy larger areas of the brain surface as compared to nonmusicians [18], and diligent piano playing over a long time results in a refined hand sensibility [19]. The hand's cortical projection normally covers a very large area in the brain, but in professional violin players, the area corresponding to the left hand (the string hand) expands further and becomes even larger, especially among those who trained in Suzuki classes in their childhood [20, 21]. The reason is probably that the left hand, through intensive training and practice, has acquired more brain resources as its skill and dexterity increased.

The creative hand's interaction with the brain is based on a delicate balance between incoming sensory (afferent) signals from the hand and outgoing (efferent) signals from the brain to the muscles in the hand. Musicians can sometimes be affected by a condition called focal dystonia, in which this balance is disrupted, resulting in difficulties in controlling the movements of individual fingers [22–28]. This is often caused by repeated and monotonous movements in the hand and wrist joint over a long period, such as many years of daily intensive training on an instrument. In such cases, the borders between the cortical representations of fingers and individual parts of the hand become diffuse, and individual fingers' cortical representations begin to overlap each other. The normal cortical hand representation can be transferred to mosaiclike patterns (Fig. 9.3). The changed mapping of the hand generates new and strange messages to the motor area, disrupting the balance in sensory motor interaction and resulting in impaired mobility and precision in the hand.

Robert Schumann is one of several musicians who were afflicted by this fairly therapy-resistant condition. His career as a concert pianist ended because of disturbances in the motor function of his right hand: he could no longer control his middle and ring fingers separately. Schumann gave quite an accurate description of what it means to suffer from focal dystonia: 'I feel myself misunderstood, especially when I have pain in my hand, and seriously speaking, it is becoming worse. I often ask "Good God, why did you do this to me?' I would have had such good use of my hand, so much music lives inside me ready to be expressed, and now I can hardly control it, with one finger flopping over the finger next to it. It's terrible and fairly painful, too"' [9].

Schumann's problem instead gave us a great composer. In a letter to his mother he wrote: 'Stop worrying about my fingers, I can compose without their help and I would hardly be happier as a travelling piano virtuoso.'

The Multisensory Brain

Anyone playing an instrument utilises several senses – sensation, vision and hearing – to achieve the final expression [29]. I remember my first piano lesson in childhood, when I was just learning to navigate among the keys. For several pupils it was a relief that middle C was usually situated at the keyhole on the piano. However, I oriented myself on the basis of the pattern in the ivory on the keys – the pattern of lines in our piano's middle C reminded me of a woolly mammoth viewed from the front. But of course there was a problem; to my father's frustration, every time I had to perform at relatives' parties where I was expected to perform on a foreign piano, the pattern in the ivory was quite different and the performance usually ended in disaster. Over time the problem resolved itself when I learned how to use the positions of the white and black keys of the keyboard as references.

Experienced piano players do not even need to lower their eyes to the keyboard when they play – the sensory feedback generated by the first touch of fingertips on the keys is enough to give the hand a sufficient keyboard orientation. This is especially obvious in blind pianists: there are several well-known brilliant jazz pianists such as Art Tatum, Ray Charles and Stevie Wonder. We know that blindness sharpens the other senses, and a refined sensibility in the finger pulps and refined hearing may well be an advantage for an improvising or accompanying jazz pianist. Learned patterns regarding rhythm, harmonies and precision in the hands are there; 'the hands know perfectly well what to do'.

For piano players, the sensory feedback from the hand is a prerequisite for adjusting the force of the fingers against the keys (keystroke dynamics). The piano is basically a percussion instrument; the tones are created by wooden hammers covered with felt hitting strings, and the power in the touch determines the strength of the tone. An organ, however, works differently: an 'all or nothing' principle. The volume remains the same regardless of how hard you strike the keys. My father, who was an organist in one of the churches in Gothenburg, complained that most organists are bad piano players with little ability to nuance their playing since they were used to hitting the keys too hard. The organ – the mother of all instruments – instead offers other fascinating options for regulating the timbre, character and strength of the tones by using a panel for modulation of the tone's character, something that requires great dexterity in the hands. Moreover, church organs are equipped with foot pedals. What a task for the brain to handle such an instrument – with both hands and both feet working at the same time! Here we can talk about music that involves enormous areas of both brain hemispheres.

The creation as well as the experience of music is based on an interaction among our senses, primarily sensation, vision and hearing. Often there is a component of synaesthesia, an overlapping various sensory functions. For many musicians, different keys have different colours and generate totally different sensory experiences and feelings. My father, who worked as an organist for more than 50 years, described with great passion and empathy what a radical change of the key of a hymn could mean to him. In August of 1986, the Church of Sweden decided to make certain changes in the classic hymn book so that some hymns in the new edition should be lowered one step, making them easier for the congregation to sing. Churchgoers had long complained that some hymns were difficult to sing because they were too high pitched. The new rule was introduced the first Sunday of Advent. My father was very worried – he felt strongly that some of the Christmas hymns completely lost all their Christmassy character when lowered from the bright shimmering D major key to the bland, patriotic C major key: the aroma of straw, candle wax and lighted candles, which he felt so intensely associated with this music, had suddenly vanished.

For several composers, the choice of key is essential to the character of the music. In a 2010 interview in a Swedish newspaper, pianist and professor Hans Pålsson at Lund University quoted Johannes Brahms, who preferred to write all his compositions in the calm and safe B flat key: 'This fat cow which I perhaps have already milked too much' [13]. He also commented on Franz Schubert's very difficult-to-play G-flat major impromptu, which a music publisher wanted to raise a step, to the much more easily played G major, in order to make the piece easier to sell. But Schubert refused: 'Like tinting an image of van Gogh in green', Hans Pålsson summarised.

The ability to experience synaesthesia, especially between hearing and vision, has been described by several musicians and composers. Jean Sibelius described strong colour experiences induced by certain tones. He described green, his favourite colour, as corresponding to a tone between E and E sharp. Similar experiences have been described by Nikolai Rimsky-Korsakov and Aleksander Scriabin, who even had plans to compose a musical piece containing colour as well as smell sensations.

The Extended Hand

The hand can be regarded as an extension of the brain, able to express a creative mind's musical ambitions. In composers, the hand translates inspired thoughts and ideas into written notes, previously with the help of a quill pen and parchment, and currently often using computer programs. Still, some composers prefer to write the

Fig. 12.8 Sixten Ehrling
conducting in the Stockholm
Concert Hall 1976 (Photo:
Jan Delden)

notes on paper by hand. A close friend and composer said that inspiration occurs in a very different way when he feels the friction of the pencil on paper as compared to when the hand is working on a sterile plastic computer keyboard. Similarly, a cartoonist can feel the creative capacity of her hands drawing directly on a paper to be quite superior as compared to seeing a computer producing animated pictures: 'There is something very personal and more expressive when you draw. It is so direct – the thoughts flow freely from the brain to the hand to the pencil to the paper', according to John Musker, one of the artists behind several of the Disney movies, 'The computer animation is perfect, but it is the very imperfection in hand-drawn films that makes them come alive' [30].

In the same way, 'the exploratory sketch' can be a very important aid as a first step in the creative process of designing buildings, rooms and environments for daily living. Arne Branzell, Swedish architect and designer, a teacher of visualisation technique at Chalmers Institute of Technology in Gothenburg, emphasised the importance of a hand-drawn conceptual sketch serving as the basis for continuing the design process using digital equipment. He feels that the silent knowledge, creativity and improvisational abilities of the hand are of utmost importance before premade computer programs become too involved in the process.

The concept of the hand as an extension of the brain can be applied to many situations where the hand holds tools and various types of instruments: using a familiar tool or playing a musical instrument may induce changes in the cortical body map so that the tool or instrument, from the brain's viewpoint, becomes part of the arm and hand. The conductor's baton is not perceived as a foreign object that he holds in his hand; it becomes part of the conductor's hand, an extension of the hand with its own projection in the brain – the conductor's baton becomes part of the conductor (Fig. 12.8).

In the same way, an experienced musician perceives his instrument as a part of his body. For the violin player, the bow becomes an extension of her arm and the violin becomes a part of the musician. The hand represents the intermediary between body and instrument – the grip and the immediate contact with the violin, together with the instrument's vibrations and auditory feedback, make the instrument a part of the body – it becomes a case of 'body ownership' (Fig. 12.9).

Fig. 12.9 The young violin
player may need a helping
hand to learn how to hold the
bow. In the experienced
violinist, the instrument
becomes a part of the body
– a natural extension of the
bow and string hand (From
Claude Verdan, *La main, cet
univers*. Fondation du Musée
de la Main, Lausanne, 1994.
© Jacques Dominique
Roullier)

The feeling of 'body ownership' when using an instrument, be it a violin or a
grand piano, is probably a prerequisite for excellence in music performance. The
famous Swedish pianist Staffan Scheja expressed it this way: 'The grand piano
becomes a part of yourself; it's a matter of magic divine moments when you get a
feeling that it is not you yourself who are playing, when your thoughts go directly
into the instrument without it being felt as a product of the hand's work.'

References

1. Berg L. Gryning över Kalahari: hur människan blev människa. Stockholm: Ordfront; 2005.
2. Henshilwood CS, d'Errico F, van Niekerk KL, Coquinot Y, Jacobs Z, Lauritzen SE, et al.
   A 100,000-year-old ochre-processing workshop at Blombos Cave, South Africa. Science. 2011;
   334(6053):219–22.
3. Balter M. Human evolution. Early start for human art? Ochre may revise timeline. Science.
   2009;323(5914):569.
4. Texier PJ, Porraz G, Parkington J, Rigaud JP, Poggenpoel C, Miller C, et al. From the Cover:
   a Howiesons Poort tradition of engraving ostrich eggshell containers dated to 60,000 years ago
   at Diepkloof Rock Shelter, South Africa. Proc Natl Acad Sci U S A. 2010;107(14):6180–5.

5. Rigaud JP. Art treasures from the ice age. Lascaux Cave. Nat Geogr. 1988;174:482–99.
6. Balter M. Origins. On the origin of art and symbolism. Science. 2009;323(5915):709–11.
7. Clottes J, Courtin J. La grotte Cosquer: peintures et gravures de la grotte engloutie. Paris: Seuil, cop; 1994.
8. Coghlan A. Cave art rescue bid creates new threats. New Scientist. 2009. p. 12.
9. Sandblom P. Creativity and disease. How illness affects literature, art and music. New York: Marion Boyers; 1997.
10. Drozdov I, Kidd M, Modlin IM. Evolution of one-handed piano compositions. J Hand Surg Am. 2008;33(5):780–6.
11. Edel T. Piano music for one hand. Bloomington: Indiana University Press; 1994.
12. Patterson DL. One handed: a guide to piano music for one hand. Westport: Greenwood Press; 1999.
13. Ekström A. Interviewing Hans Pålsson about Beethoven Sydsvenska Dagbladet (Swedish newspaper). 20 Feb 2010. p. 2.
14. Schlaug G, Jancke L, Huang Y, Staiger JF, Steinmetz H. Increased corpus callosum size in musicians. Neuropsychologia. 1995;33(8):1047–55.
15. Gaser C, Schlaug G. Brain structures differ between musicians and non-musicians. J Neurosci. 2003;23(27):9240–5.
16. Hutchinson S, Lee LH, Gaab N, Schlaug G. Cerebellar volume of musicians. Cereb Cortex. 2003;13(9):943–9.
17. Bengtsson SL, Nagy Z, Skare S, Forsman L, Forssberg H, Ullen F. Extensive piano practicing has regionally specific effects on white matter development. Nat Neurosci. 2005;8(9):1148–50.
18. Pantev C, Engelien A, Candia V, Elbert T. Representational cortex in musicians. Plastic alterations in response to musical practice. Ann N Y Acad Sci. 2001;930:300–14.
19. Ragert P, Schmidt A, Altenmuller E, Dinse HR. Superior tactile performance and learning in professional pianists: evidence for meta-plasticity in musicians. Eur J Neurosci. 2004;19(2):473–8.
20. Elbert T, Pantev C, Wienbruch C, Rockstroh B, Taub E. Increased cortical representation of the fingers of the left hand in string players. Science. 1995;270(5234):305–7.
21. Hashimoto I, Suzuki A, Kimura T, Iguchi Y, Tanosaki M, Takino R, et al. Is there training-dependent reorganization of digit representations in area 3b of string players? Clin Neurophysiol. 2004;115(2):435–47.
22. Bara-Jimenez W, Catalan MJ, Hallett M, Gerloff C. Abnormal somatosensory homunculus in dystonia of the hand. Ann Neurol. 1998;44(5):828–31.
23. Rosenkranz K, Butler K, Williamon A, Cordivari C, Lees AJ, Rothwell JC. Sensorimotor reorganization by proprioceptive training in musician's dystonia and writer's cramp. Neurology. 2008;70(4):304–15.
24. McKenzie AL, Nagarajan SS, Roberts TP, Merzenich MM, Byl NN. Somatosensory representation of the digits and clinical performance in patients with focal hand dystonia. Am J Phys Med Rehabil. 2003;82(10):737–49.
25. Levy LM, Hallett M. Impaired brain GABA in focal dystonia. Ann Neurol. 2002;51(1):93–101.
26. Stinear CM, Byblow WD. Impaired modulation of intracortical inhibition in focal hand dystonia. Cereb Cortex. 2004;14(5):555–61.
27. Winspur I, Wynn Parry CB. The musicians hand. London: Martin Dunitz; 1998.
28. Tubiana R, Amadio PC. Medical problems of the instrumentalist musician. London: Martin Dunitz; 2000.
29. Zatorre RJ, Chen JL, Penhune VB. When the brain plays music: auditory-motor interactions in music perception and production. Nat Rev Neurosci. 2007;8(7):547–58.
30. Geistrand M. Interviewing J Musker. Sydsvenska Dagbladet (Swedish newspaper). 1 Feb 2009.

Chapter 13
Right Hand or Left Hand?

Abstract Though the aetiology of left- and right-handedness is not well understood, it is believed genetics may play an important role. In our society most industrial products, electronic equipment, mechanical tools and kitchen utensils are produced for right-handed people, something that may cause considerable problems for the left-handed. Left-handedness has often been associated with creativity, musicality and artistic talents, and many of our most famous painters and successful people in culture and society are lefties. From the linguistic point of view, the words corresponding to *right* and *left* represent good and evil, and in many cultures the symbolic roles of the right and left hands are quite different. The dominance of right-handedness or 'right behaviour' might have evolved very early in evolution and seems to be present not only in humans but also in several animal species.

The terms right and left appear in many various contexts, in chemistry, politics and biology. Molecules can be left- or right-oriented, that is, two different types that relate to each other as the right hand relates to the left hand – like mirror images. In the 1860s, Louis Pasteur noted that a certain type of tartaric acid crystals rotated polarised light clockwise while a mirror variant instead rotated the light anticlockwise. Even people can be left or right oriented, but in that case we're talking about differences in political opinion.

Being right- or left-handed is something quite different [1–5]. Why do most people have the right hand as their 'dominant' hand – the hand that intuitively feels like the most natural to use and that possesses the greatest precision and power – and a minority feel the left hand is the dominant one?

The left-handed literature researcher Niklas Schiöler at Lund University, Sweden, has reflected on all aspects of left-handed people's problems with humour and great insight, as he precisely describes several of the everyday problems that left-handers may encounter [6]. Being left-handed is hardly something that people notice. At most it is sometimes regarded as an interesting trait since many people feel that left-handedness is associated with positive properties like creativity, musical and artistic talent. However, it has not always been this way. There are many left-handed elderly people who have bitter memories of rude teachers who did not accept the left hand as the dominant one. There are several stories about the anxiety that followed forced relearning from the left to the right hand in their early years. Benjamin Franklin, who was left-handed, described how he was corrected and even beaten each time he used his left hand at school.

Historically, the left hand has always represented something negative and unsuccessful. In the Bible the left hand was contaminated by evil forces. Left-handedness was regarded a malformation often associated with evil and criminality. In some cultures, the left-handed were regarded the devil's lackeys. In folklore, the left-handed were closely associated with the devil, and left-handedness in women could be proof of witchcraft and even a reason to be burnt at the stake.

In several cultures, the right hand represents purity and truth, while the left hand is used to manage hygiene during visits to the toilet. Niklas Schiöler describes how he once was invited to a Muslim wedding in Kashmir and was unaware of the strict right-left rules. Schiöler, who was a fairly non-prominent guest, had to eat with the servants in a room on the bottom floor where everybody was sitting on the floor. When the meal was served, everybody expected the foreign guest to be the first to start the meal. The left-handed Schiöler stretched out his hand – unfortunately the left hand – to pick some pieces of food. Immediately the whole group stopped talking and looked at each other horrified, and the guest immediately understood his mistake. The left hand is okay when using the toilet, but to use it at the dinner table is unforgivable. Many explanations were needed to salvage the situation.

In language as well, there are several indications of how sensitive left-right terminology can be. During the time of Emperor Tiberius, the Latin word *dexter* (right) represented something bringing luck and welcoming, while *sinister* (left) arose from *sinistrum* meaning evil. Individuals born left-handed were regarded as offensive. *Dexterity* means adroitness and precision in the hand. *Right hand* indicates the correct and appropriate hand, while left is derived from the Anglo-Saxon *lyft* meaning weak and broken down. *Left-handed business* means illegal and immoral business. In the German language, *link* (left) stands for incorrect. While the French *droit* means right, correct, straight and honest, *gauche* represents left, clumsy or awkward.

Being Left-Handed

In our part of the world and in our time, the left hand is, thankfully, equally appreciated as the right hand. I discovered early on that both of my sons were left-handed, which my wife and I found somewhat strange since we are both right-handed. But we felt that our sons' left-handedness perhaps after all was fairly natural – they are both creative and musically talented, and there seems to be a general belief that left-handedness is linked to creativity and musicality. No one in the family paid much attention to our sons' left-handedness until the younger one came home from school 1 day beaming with joy, telling us that he no longer had to participate in sewing class – the teacher had given up after hopeless attempts to teach him crocheting and knitting. His mirror-reversed motor functions were impossible to correlate with the teacher's instructions. The fact that he could not use regular scissors also gave him advantages at school and at home. His inability to handle a standard potato peeler automatically relieved him from several kitchen activities, until the day we stumbled upon a special shop for left-handed people. Here we found everything that a parent of left-handed children could want: specially made and modified potato peelers and special scissors made for the left-handed, plus many other things that to this day disqualify our left-handed children from special treatment at home and at school.

Society is indeed not designed for left-handed people, at least not with respect to all the everyday items and tools that are natural parts of daily living. In the days of the old farming society, the left-handed must have had many problems with several activities, for instance, hay-making, since the handles of the scythe are always mounted for right-handed use. To have a left-handed man on the team making hay with a reversed stroke would be very dangerous when several men are working at the same time side by side. Chris McManus quotes in his book *Right hand, Left Hand* a very descriptive section from Tolstoy's Anna Karenina [1]:

> …the peasants came into sight, some with their coats on, some in their shirts, following one behind another in a long string, each swinging his scythe in his own manner. Levin counted forty-two of them…He heard nothing save the swish of the knives, saw…the crescent curve of the cut grass, the grass and flower-heads slowly and rhythmically falling about the blade of (the) scythe…On the short rows the mowers bunched together…their scythes ringing when touched.

Levin realised how dangerous a tool a sharpened scythe could be, 'sharp as a razor blade': 'A curved piece of steel with a carefully honed edge, several feet in length, swung in a huge arc around the mower. To get one's feet in the way was to risk serious injury. With a team of men scything their way across a field it was vital that they all be synchronized. To have one man doing everything back to front – left-handed, in other words – would be to risk disaster.'

McManus describes several left-handed problems. For the left-handed, the natural turning movement goes anticlockwise, while the control buttons on ovens, stereos and all kinds of electronics are turned clockwise. Serving wine at a meal from the right side, according to strict etiquette, is very risky for the left-handed – filling glasses with precision is very difficult. Even at the dinner table, the left-handed can encounter problems if the neighbour on the left is right-handed: enjoying soup with a spoon in the left hand can cause many problems since the lefty's elbow will easily collide with his neighbour's right elbow. Even when handling a knife and fork, the left-hander has to cross over the right hand to reach the knife, which always is positioned to the right of the plate.

Among other well-known problems are handling screwdrivers, can openers, corkscrews and traditional types of pencil sharpeners. Using cutting and grinding tools or circular saws can be very risky for the left-handed. Zippers and buttons, as a rule, are constructed for right-hand use, as are coffee machines, musical instruments, cameras and slot machines. The wristwatch is placed on left wrist so that the crown can be reached by the right hand. Computer games usually require dexterity in the right hand since most of the commands are handled from the right side of the keyboard. On a laptop the 'Enter' key is placed to the right in an uncomfortable position for the left-handed. And when the left-hander enters or leaves a room, the door handle is placed in a very inconvenient position.

Left-handedness can certainly be more or less prominent. While some people are exclusively left-handed, others seem to be able to use both hands interchangeably. Moreover, the brain seems to very rapidly be able to decide which of the hands is 'dominant' in various situations. A right-handed person uses the right hand to take a Coca-Cola bottle from the refrigerator but switches hands to unscrew the cap, holding the bottle in the left hand while using the right for the more qualified task of unscrewing the cap.

Famous Lefties

Those who are left-handed are in fact members of an exclusive group of prominent people. Many famous politicians, philosophers, musicians and artists all have dominant left hands. Left-handed people are also well-represented among artists, sculptors, architects and musicians. Among left-handed artists are Leonardo da Vinci, Michelangelo, Rafael, Rubens, Albrecht Dürer and Paul Klee. In music there are Sergei Rachmaninov, Benjamin Britten, Niccolò Paganini, Robert Schumann and Maurice Ravel, and Paul McCartney as well as Jimi Hendrix. Among left-handed writers are Johann Wolfgang Goethe, Mark Twain and Friedrich Nietzsche. Many American presidents have been left-handed, including Gerald Ford, Ronald Reagan, George Bush Senior, Bill Clinton and Barack Obama. So even if it is not scientifically proven that left-handedness is linked to creativity and success, at least it does not seem to be a disadvantage [1, 6].

It seems left-handers also have a price to pay. Left-handed people have a greater tendency to suffer from allergies, asthma, dyslexia and diabetes [1] than righties. Intestinal diseases like ulcerative colitis and Crohn's disease also seem more common among left-handed people. On the other hand, left-handed males can, statistically, look forward to more well-paid occupations and a higher income as compared to their right-handed brothers [7, 8].

How Long Has Left-Handedness Existed?

Left-handedness has probably existed as long as humanity has existed. In early history, left-handed people were often associated with bad properties and dark events, sometimes even with murder. The Bible often expresses mixed feelings for left-handed people, who were also regarded as brave and effective warriors. The Book of Judges describes how the Benjamites were involved in a bitter war with the Israelites and how the Benjamites mobilised 26,000 armed men. Among them, chosen from the people of Gibea, were 700 left-handed elite soldiers, 'each of whom could sling a stone at a hair and not miss' (Judges 20:15–16). Schiöler has described how left-handed people, during peacetime, lived with the cattle and were generally made fun of but how in battle they were put on the front line: 'they should be the first to kill and the first to be killed' [6].

How Common Is It to Be Left-Handed?

No one doubts that right-handedness is considerably more common than left-handedness; we can just look around among our closest friends. But how common is it to be left-handed? The issue is not simple to clarify since many people use the right and left hands to the same extent in various types of activities, while others have a preference for one or the other hand in various types of gripping functions. Moreover, the proportions of right- and left-handedness vary somewhat with the age and place of residence [1]. Nigel Sadler at Vestry House Museum questioned 3,000 schoolchildren in the Waltham Forest District in northern London about which hand they used in various activities. Those who used the left hand in half or more of their occupational tasks were considered left-handed. He found that a little more than 10 % were left-handed. He also found that left-handedness is somewhat more common among boys, 11.6 %, as compared to 8.6 % among girls. These results are consistent with other studies that usually find five left-handed males as compared to four left-handed females.

How Early in Life Does Left-Handedness Occur and What Is the Reason?

How early in life is it possible to estimate whether a child is left- or right-handed? These are difficult and controversial questions and we lack definite answers. But there are several speculations and interesting observations. It is often said that right-handedness or left-handedness does not appear until the second year of life, but there are also observations indicating that the handedness indeed appears very early on, even in the embryonic stage. McManus reports a study carried out by Peter Hepper at Queen's University of Belfast, using an ultrasound technique to investigate the embryo's activity early during pregnancy. He found that during the 10th embryonic week the right arm moved much more than the left arm and in the 12th embryonic week he could see the embryo sucking the thumb – in more than 90 % the right thumb.

But what is the aetiology of left-handedness? There are various theories. It has been speculated that an abnormally high influence of testosterone during the embryonic stage might delay the development of the left hemisphere, which controls the right hand. Difficult deliveries with increased loading of the brain have also been proposed as a plausible reason. But it is probably some kind of genetic influence that determines whether the child will be right- or left-handed. Researchers have different opinions regarding the relative importance of genetic factors and environmental factors, even if no one today denies that left-handedness can be a family trait. But it is not easy to sort out how the hereditary disposition is transferred over generations. In 2004 Phil Bryden and Chris McManus critically analysed all available scientific reports on handedness. Seventy thousand children were studied. The results showed that when both parents were right-handed, the chance of having a left-handed child was 9.5 %; if one parent was right-handed and the other left-handed, the probability was 19.5 %, and if both parents are left-handed, the probability increased to 26 % [1].

Another example of left-handedness as a family trait is the British royal family, where left-handedness has been transferred in a clear descending line: George II, George IV, George VI, Victoria, the queen's mother Elizabeth and Prince Charles.

Thus left-handedness is obviously associated with some kind of family trait, but must it still have a genetic base? Several properties that are inherited over generations may have a component of cultural and environmental influence, and with regards to left-handedness, this may sometimes be the case. Often there is a strong social advantage of right-handedness rather than left-handedness. This makes it difficult to rule out genetic factors from influencing left-handedness.

However, the advocates for genetic factors have a problem that is obvious in identical twins. Both twins originated from the same egg and therefore have identical genes, yet they do not always share the property of right- or left-handedness – in about one out of five pairs, one twin is right-handed and the other one is left-handed [1].

Thus there are many mysteries regarding the aetiology of left-handedness. Ultimately it is a matter of redistribution of work between the brain hemispheres. The left hemisphere controls movements in the right hand, while the right hemisphere controls movements in the left hand. But which factors determine which role each hemisphere is to have is not known in detail. Perhaps there is some crucial decisive genetic component, but as yet no one has been able to isolate a DNA sequence that is systematically different in right-handed as compared to left-handed people [1].

Was Lucy Right- or Left-Handed?

No one can say for sure how long left-handedness has existed in the development of humankind. However, there are signs of a dominant right hand in our early ancestors. Palaeoanthropologists argue that 3.2 million years ago Lucy (*Australopithecus afarensis*) was right-handed. Fossils from skulls of apes show obvious injuries on the left side of their skull, probably generated by branches or bone structures from antelopes that were used as weapons by right-handed hominins. It is also argued that *Homo habilis*, the first known hominin to make tools, showed obvious signs of a dominant right hand. *Homo habilis* used toothpicks, and judging from worn marks on the teeth, such toothpicks were held in the right hand [1]. Finds of stone tools also indicate right-handedness: the anthropologist Nicholas Toth investigated remnants of stone tools in Koobi Fora, Kenya, which have been dated to 1.8 million

Fig. 13.1 Right-handed scribe on a 2,000-year-old wall painting in Pompeii. The pen is held against the scribe's lips in a gesture of attention and thoughtfulness

years ago. When making stone tools, it is natural for a right-handed person to hold the stone in the left hand and to knock off flakes with a mallet stone in the right hand. According to Toth, the processing of stone tools is typical for work performed by right-handed individuals [9–11].

Also, later in history there are indications that our ancestors were right-handed. Negative handprints in Patagonia, about 9,000 years old, show almost exclusively the contours of left hands. Probably the artist was right-handed, using the right hand to fill the mouth with colour pigments that were then sprayed around the left hand placed on the cliff wall.

The Biblical Eve in Eden is portrayed in many illustrations with the fateful fruit in her right hand. In pictures from the time of the Pharaohs, illustrating the earliest forms of postal services, the documents were written by the right hand, and on paintings from the ruins of Pompeii from the first century AD, the dominant right hand is obvious. A well-known wall painting from Pompeii depicts a scribe, probably a woman, attentively listening to a letter or a document being dictated. She holds the pen in her right hand, its tip resting against her lower lip in a gesture of thoughtfulness and reflection (Fig. 13.1).

Theories About the Origin of Right-Handedness

Thus, dominant right hands can be traced far back in time, but how did right-hand dominance arise? Does right-handedness express some kind of natural selection of right-handed individuals during evolution, or is it a result of specialisation of both hemispheres evolving over an extended time period? There was once a theory that there was indeed a kind of natural selection, since our ancestors used their left hands in battle to protect their hearts with shields: thus it was necessary to hold the sword in the right hand. Warriors who did so would have a greater survival rate than those who held the sword in their left hands, leaving their heart more unprotected. The right-handed would therefore have a greater chance of survival and thus preserve their genes for the next generation.

According to recent theories, right-hand dominance and the respective roles of both hemispheres might have arisen hundreds of million years ago when the first vertebrates appeared on land. And right-hand dominance is perhaps not at all a specific human characteristic; according to MacNeilage it can be demonstrated in most animal species [12]. Right-handedness and left-handedness reflect the respective functions of both brain hemispheres. It is well known that the hemispheres have distinctly different roles. The left controls language and regulates the movements of the right hand. It analyses and classifies and is involved in mathematical calculations. The right hemisphere, on the contrary, is more artistic and emotional and regulates the movements of the left hand. MacNeilage and his associates argue that such a distribution of roles between the hemispheres for some reason evolved very early in the development of vertebrates.

A 'right behaviour' is obvious in most animal species, for instance, fishes, amphibians, birds and whales. Catching prey is an example of a well-established routine behaviour, controlled by the left hemisphere. In laboratory experiments, frogs prefer catching prey that is positioned on their right, and at the Alaska Fisheries Science Center in Seattle, researchers observed that whales have more scratches on their right part of their cheek area. Also baboons, macaque monkeys and chimpanzees show an obvious 'right behaviour'. William Hopkins at Yerkes National Primate Center in Atlanta has studied the behaviour of laboratory monkeys handling tools that require activities using both hands [12]. If an opened jar containing honey was placed in front of the monkey, the animal usually preferred to scrape out honey with one of the right hand's fingers.

Perhaps right-handedness appeared very far back in evolution, before the ape line and the human line were separated in the developmental tree about 6–7 million years ago. But the very reason for the specialisation of both hemispheres and the dominant right behaviour is still a mystery that may never be solved.

References

1. McManus C. Right hand, left hand: theories of asymmetry in brains, bodies, atoms and cultures. London: Weidenfeld and Nicolson; 2002.
2. Coren S. Left-handness: behavioral implications and anomalies. Amsterdam: North-Holland; 1990.

3. Diane P. The left-handers handbook. Stourbridge: Robinswood Press; 1998.
4. Annett M. Handedness and brain asymmetry: the right shift theory. Hove: Psychology; 2002.
5. Langford S. The left-handed book. London: Panther; 1984.
6. Schiöler N. Avig eller rätt: en vänsterhänt betraktelse (In Swedish). Stockholm: Carlsson; 2007.
7. Ruebeck CS, Harrington Jr JE, Moffitt R. Handedness and earnings. Laterality. 2007;12(2):101–20.
8. Denny K, O'Sullivan V. The economic consequences of being left-handed. J Hum Resour. 2007;42:353–74.
9. Ambrose SH. Paleolithic technology and human evolution. Science. 2001;291(5509):1748–53.
10. Schick KD, Toth N. Making silent stones speak: human evolution and the dawn of technology. London: Weidenfeld & Nicolson; 1993.
11. Toth N. Archeological evidence for preferential right-handedness in the lower and middle Pleistocene, and its possible implications. J Hum Evol. 1985;14:607–14.
12. MacNeilage PF, Rogers LJ, Vallortigara G. Origins of the left & right brain. Sci Am. 2009; 301(1):60–7.

Chapter 14
Losing a Hand

Abstract Amputation of a hand results in profound synaptic reorganisations in the brain cortex. The cortical representational area of the amputated hand is at first 'silent' but rapidly becomes invaded by adjacent cortical areas. Cortical reorganisations can result in *phantom sensation*, a feeling that the lost hand is still there. *Phantom sensation* can sometimes be combined with severe *phantom pain*. Mirror treatment, based on an illusion that the lost hand is still attached, may be an effective way of treating severe phantom pain. Amputated fingers or hands can often be microsurgically replanted (reattached to the body). When a replanted hand is reinnervated, it resumes its correct position in the brain cortex. The salamander is unique among animals since it can spontaneously regenerate an amputated extremity. It has been suggested that the mechanism behind this phenomenon is based on an interaction of stem cell-like cells at the amputation level interacting with a specific Schwann cell-produced protein (nAG). If these mechanisms could 1 day be applied to humans, it would open up a totally new landscape for treating amputations.

Losing a hand is catastrophic, an event that may have lifelong consequences. There are several reasons for accidental hand amputation. I have had patients who walked into the rotating propeller of a small airplane or tilted a motor-driven lawnmower or were not sufficiently cautious while working with a saw or handling fireworks on New Year's Eve. Sometimes, an arm can be amputated by choice, because of tumours.

In addition, children can be born without a hand or an arm as a type of congenital malformation. Whatever the reason, the consequences can be quite severe. Many patients perceive the amputation as a mutilation of both body and identity, a psychological trauma that can be very hard to accept and that can create a personal crisis. The amputation may make it difficult to continue one's occupation, and major adjustments in daily life may be necessary. It may also be a cosmetic problem that can be difficult to handle emotionally.

G. Lundborg, *The Hand and the Brain*,
DOI 10.1007/978-1-4471-5334-4_14, © Springer-Verlag London 2014

Phantom Sensation and Phantom Pain

A common complication after an amputation is a phantom sensation, a feeling that
the amputated hand is still in place [1–5]. The phantom sensation may be enormously painful. The experience is often frightening and quite obvious, many times
with a feeling that individual fingers are moving or maybe locked in painful contractures, sometimes so severely it feels as if the nails are being forced into the
palm. Some patients feel that they can move the fingers of the phantom hand, while
others feel as if it is totally paralysed. Often phantom sensation is linked to a severe
phantom pain that can be burning, cramp-like or stabbing.

What Happens in the Brain After a Hand Amputation?

Phantom sensation and phantom pain are based on the very prominent functional
reorganisations in the brain cortex that follow an amputation [6, 7]. Amputation of
a hand or an arm results in rapid changes in the cortical body map [8–11]. The cortical representation of the hand is normally quite large in the sensory and motor cortices, but after an amputation there is no longer a sensory inflow from the hand. The
hand projection in the cortical body map suddenly becomes silent, and this area
becomes 'vacant' and unused: there is a 'black hole' in the brain. The brain is ratio-
nal and does not accept silent areas. Within minutes and hours there is a functional
reorganisation in the synaptic connections. The adjacent cortical projectional areas
corresponding to the face and the residual part of the forearm now expand over what
previously belonged to the hand, and the nerve cells in this area now make new
functional connections. Among several consequences of amputation of an arm is a
change in the face's sensory functions. After only 24 h, there may be obvious indications of functional reorganisations in the somatosensory cortex, where the face
representation expands over the previous hand-arm representation [9, 12–15]. The
result is often a 'mapping' of the hand in the face so that touching various parts of
the face results in sensations in various parts of the non-existent phantom arm [3].

After a hand amputation, there is often a 'mapping' of the lost hand on the skin
of the residual part of the forearm [3, 15–18]. This is presumably a consequence of
the cortical reorganisations that occur after the amputation, including the expansion
of the adjacent forearm representation, just as the amputation of a finger results in
the expansion of the adjacent cortical representations in the somatosensory cortex
of nearby fingers (Fig. 14.1). Following the amputation of an index finger, the cortical projections of the thumb and long fingers expand so that these fingers take over
the area that previously belonged to the index finger [6, 9, 19–22].

The functional reorganisation that occurs in the brain cortex following amputation is the primary reason for the feeling that the amputated body part is still intact.
Sometimes the phantom sensation can be linked to phantom *pain*, which can sometimes be extreme and almost impossible to endure [14, 23–25]. It is well known that

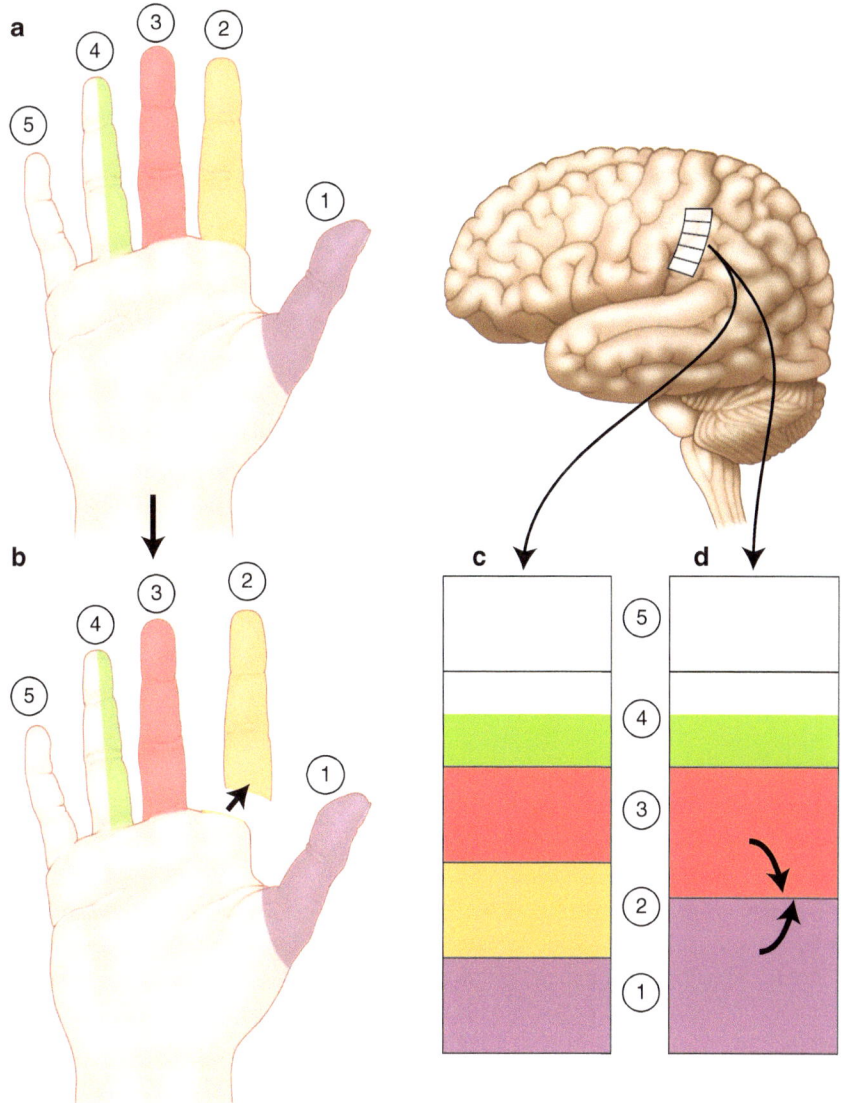

Fig. 14.1 Functional reorganisation occurring in the sensory cortex following amputation of an index finger. When a finger is amputated (**a, b**), a corresponding 'silent' area occurs that is rapidly occupied by the adjacent representational areas corresponding to the thumb and middle finger (**c, d**)

the intensity of phantom pain is in proportion to the extent of the synaptic reorganisations in the brain: the more extensive the reorganisation, the more the pain [23, 26]. Usually phantom limbs are perceived in the location previously occupied by the intact limb, but sometimes they can retract inside the stump, a

Fig. 14.2 Edvard Munch. Self-Portrait. Between the Clock and the Bed, 1940–1943. Munch lost his left index finger in a shooting accident. The missing finger is painted in *green* to illustrate that it in fact is a 'phantom finger' with constant pain. Oil on canvas, 120.5 × 149.5 cm (© Munch-Museet/Munch-Ellingsen gruppen/BUS 2013)

phenomenon referred to as 'telescoping'. Telescoping is relevant from a clinical point of view, as it tends to be related to increasing levels of phantom pain [27].

The Norwegian artist Edvard Munch dramatically depicted the intensity of phantom pain in a self-portrait after he lost a finger (Fig. 14.2). Munch had a traumatic relationship with Tulla Larsen, the rich daughter of a wine merchant. When their relationship ended in 1902 after an unsuccessful attempt at reconciliation, Munch, in despair, shot himself in his right hand so that his index finger was amputated. In a self-portrait, Munch expressed the despair, vulnerability and defencelessness that he felt after the episode. He was affected by extremely painful phantom sensations including a component of phantom pain, something that, in his self-portrait, he illustrated by painting the non-existing index finger green in its normal position.

Mirror Training: A Way to Treat Phantom Pain

The extent of phantom sensation and phantom pain is proportional to functional changes of the amputated body part's representation in the brain. But if the modified mapping of the amputated body part could somehow be normalised, perhaps the

Fig. 14.3 'Mirror treatment' of severe phantom pain after amputation of the left arm and hand. A reflected mirror image of the intact right hand in the position of a normal left hand gives an illusion of the left hand being reattached to the body in its correct position

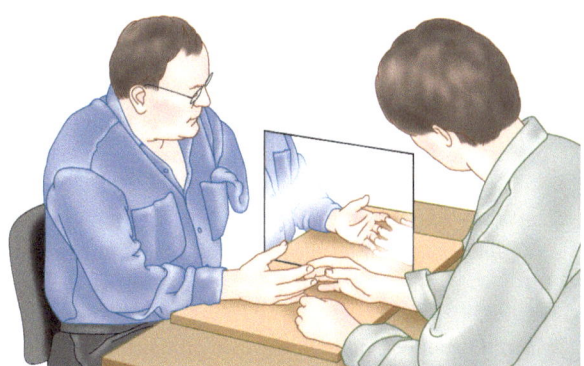

phantom pain would disappear. Such a treatment principle was first described by Ramachandran [3, 28]. The strategy is to create an optical illusion of the lost hand so that the brain gets an impression that the hand is in its normal position. This is done by placing a mirror in front of the amputee in a somewhat oblique position (Fig. 14.3). By creating a reflection of the non-amputated hand in the position of the lost hand, an illusion is created; when the patient looks in the mirror observing the mirror image, he sees a normal hand in the position of the lost hand. If the normal hand is activated so that the corresponding movements occur in the mirror picture, the illusion becomes even stronger; the visual impression is that the lost hand is not only in place but that it even functions. The visual impression of the hand being reattached may presumably induce a normalisation of the modified cortical hand-arm representation, resulting in relief or the disappearance of the phantom pain. The treatment, in its original or a modified form, has to be repeated several times but can ultimately have a positive effect [5, 29, 30].

Thus, creating an optical illusion that the amputated hand is in its usual position may be a way to treat phantom pain. But the best way to deal with the potential problem is, of course, to perform a surgical reattachment of the hand as soon as possible after amputation – a replantation.

Hand Replantation

The concept of reattaching amputated extremities can be traced back to the Renaissance and even earlier. In several Christian religious paintings, there are examples how severed hands and arms were miraculously reattached to the body [31]. In a painting by Giovanni di Paolo from the fifteenth century, a child's right arm was amputated by a wolf. A miraculous replantation was performed by Saint Clara of Assisi who first killed the wolf and then reattached the arm after it had been retrieved from another friendlier wolf. Other examples include the replantation of a thumb, depicted by an anonymous artist, the master from Novara, in a fresco painting from the fifteenth century in a northern Italian Benedictine monastery.

The painting shows how Saint Julius from Novara replanted the thumb with the help of the cross sign: 'he made the cross-sign and the hand became again as usual' [31].

Saint Julius' miraculous ability was envied; replantation is much more difficult for today's surgeons. The last four decades have seen dramatic improvement in the replantation process. In the 1960s, it was argued that an amputated finger or hand was hopelessly lost, but due to evolving microsurgical techniques in the 1970s, successful replantation surgery is now possible in many cases [32].

Replantation is a very tedious and time-consuming surgical procedure: All tissues – the blood vessels, bone structures, skin, tendons and nerves have to be repaired. Blood vessels are repaired using microsurgical techniques, a procedure performed under microscopic magnification requiring considerable technical skill as the diameter of the blood vessels may be only tenths of a millimetre.

After replantation, nerve fibres have to regenerate and grow distally to reinnervate sensory receptors and the hand's muscles. Following replantation an adult can usually recover the ability to feel temperature as well as pain and pressure, but the discriminative ability to recognise the shape of small items and the texture of surfaces is more difficult. The reason is that a large number of nerve fibres are misdirected so that, for instance, the nerve fibres belonging to the index finger instead regenerate into the thumb and vice versa [6]. The result is that the cortical projection of the hand is completely reorganised, 'the hand speaks a new language to the brain' and the brain has to learn this new language (cf. Fig. 9.4).

When a replanted hand recovers sensory and motor functions, sensory signals from the hand are again transmitted to the somatosensory cortex, and the muscles of the hand receive instructions from the motor cortex. Once again, the hand can be perceived as a true part of the body, the 'body ownership' is re-established, and a functional magnetic resonance image (fMRI) shows that the hand resumes its original position in the brain cortex [33].

The tissues in the hand show a varying vulnerability to the lost blood supply. Muscle tissue is most vulnerable and can be damaged after only 4–5 h of lost blood flow. In contrast, the fingers' tissues are more tolerant and can survive lost blood flow for at least 24 h. The critical time can be extended if the hand or finger to be replanted is cooled down to a few degrees above freezing. Therefore, it is recommended that an amputated body part should be kept in a moist dressing inside a closed plastic bag inside another plastic bag of ice during transport to the hospital.

There are other ways to involuntarily keep a finger cooled down. A fisherman happened to cut off his finger with his big fillet knife while fishing on a lake. The finger fell into the water and he had to go to the nearest hospital without it. But in the hospital someone realised that the finger, resting on the lake bottom where the temperature is about 6–7 °C, might survive for a long time and that it was worth trying to find it. A diver was sent out, he found the finger on the bottom of the lake, and a successful replantation was performed.

There is a story about a Danish policeman who came to help when a person had accidently amputated one of his thumbs. The policeman resolutely put the thumb in his own mouth during the transport to the hospital – he believed it was an advantage to keep the thumb in such a biological environment. It is said that he was awarded a medal and a lot of praise, but in fact the policeman presented a good example of what should *not* be done: in the oral cavity the temperature corresponds to body temperature and, besides, the oral cavity is enormously rich in bacteria.

The Salamander Hand

A dream for the future would be to find a way to make an amputated finger or hand regrow and reconstruct itself like a salamander's (Fig. 14.4). The arm of a salamander is small, but otherwise it does not differ much anatomically from a human arm. Like a human arm, it contains bone, muscles, tendons, nerves, ligaments and blood vessels. But the salamander arm is unique because a new extremity can grow from the residual stump that is left after amputation [34, 35]. This can be repeated several times; a new arm is regenerated after each new amputation. It seems as if the salamander in some way has kept the ability to generate new body parts, a recapitulation of what happened during the early embryonic stage. Besides being able to regenerate the arm, there is also some kind of understanding of the point of amputation. If the hand is amputated, a new hand grows, but if the amputation occurs at the upper or forearm level, the process first starts with regeneration of a new upper or forearm before a new hand is regenerated more distally at the end of the regenerated arm.

The mechanisms behind the salamander's new arm regrowth are of course of great interest. If we could understand the biological mechanisms behind this phenomenon, perhaps it would be possible sometime in the future to imitate the salamander – using the same mechanisms to regenerate an amputated human finger,

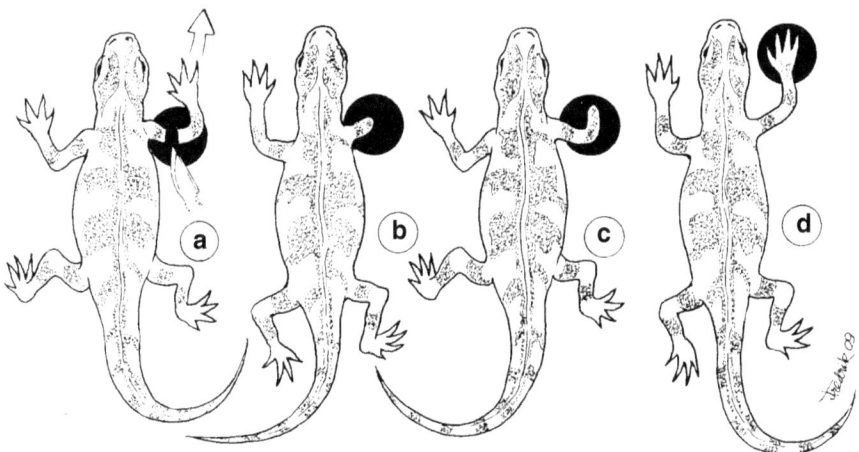

Fig. 14.4 The salamander hand. After amputation of an arm, there is a spontaneous regeneration of a new extremity including a hand (**a–d**) (Illustration: Fredrik Johansson)

hand or arm. When a human loses a finger, hand or arm, the remaining stump heals according to well-known mechanisms for wound healing, including ingrowth of connective tissue cells into the wound area, producing a scar. If the wound surface is not too large, epidermal cells can migrate from the borders of the wound to cover the scar. Even in humans evolution seems to have preserved a small part of the salamander's capacity: an amputated fingertip can heal spontaneously and become covered by epidermis, thereby being reshaped into a fairly normal appearance. Amputation of the most distal parts of fingers, therefore, is usually treated without surgery but rather with repeated changing of dressings until the fingertip has healed and spontaneously reconstructed itself.

What is the basis for the salamander's special ability to regenerate a whole new extremity after amputation? And would it be possible 1 day to apply knowledge of the salamander's healing capacity to humans? Some scientists believe that this could very well be a reality [34]. In recent years, much attention has been paid to the mechanisms behind the salamander's regeneration capacity [34–39]. Connective tissue cells in the vicinity invade the wound area from the sides and form not a scar, but a blastema consisting of cells resembling those that are present in the salamander during the embryonic stage when extremities are being formed. The cells in the blastema have a stem cell-like character and can regenerate a new extremity. The process has been interpreted as a recapitulation of the formation of extremities that took place during the early embryonic stage, and it has been argued that the same genetic programme is active in both situations. The salamander has simply switched on the genetic programme that was active during the embryonic stage and still exists, hidden in the genetic instruction manual of the cells even in the adult stage.

Interestingly, a salamander's arm can only regenerate if nerve fibres are still present in the amputated extremity, regenerating distally in the stump [36]. Thus, the regeneration of the extremity is based on some kind of interaction between the regenerating nerve fibres and the blastema at the amputation level. If nerves are injured proximally in the extremity and if the regeneration of nerve fibres is inhibited, there is no regeneration of the amputated extremity.

The regeneration of an amputated salamander extremity is regulated by complex biological mechanisms where the influence of several growth factors and specific proteins play a major role. In the search for the molecular basis for the salamander's regeneration ability, a specific protein has been identified, nAG, which is produced in the Schwann cells – the cells that form a sheath surrounding the nerve fibres in the nerve trunks in the arm. Later, the nAG protein can also be found in the epidermis layer that grows to cover the blastema. The interaction between the nAG protein and the blastema is necessary to induce regeneration of a new extremity with all its components.

Perhaps 1 day it will be possible to imitate a salamander and to regenerate an amputated hand, a finger or an arm. The principle would be to restart the genetic programme that regulated the generation of our extremities in the embryonic stage. If so, a whole new panorama would open up regarding treatment of amputation injuries and possibly for treatment of children who are born missing a limb.

References

1. Giummarra MJ, Georgiou-Karistianis N, Nicholls ME, Gibson SJ, Chou M, Bradshaw JL. Corporeal awareness and proprioceptive sense of the phantom. Br J Psychol. 2010;101(Pt 4): 791–808.
2. Ramachandran VS, Rogers-Ramachandran D, Stewart M. Perceptual correlates of massive cortical reorganization. Science. 1992;258(5085):1159–60.
3. Ramachandran VS, Blakeslee S. Phantoms in the brain: human nature and the architecture of the mind. London: Fourth Estate; 1999.
4. Yang TT, Gallen C, Schwartz B, Bloom FE, Ramachandran VS, Cobb S. Sensory maps in the human brain. Nature. 1994;368(6472):592–3.
5. Ramachandran VS, Altschuler EL. The use of visual feedback, in particular mirror visual feedback, in restoring brain function. Brain. 2009;132(Pt 7):1693–710.
6. Lundborg G. Nerve injury and repair. Regeneration, reconstruction and cortical remodelling. 2nd ed. Philadelphia: Elsevier; 2004.
7. Mercier C, Reilly KT, Vargas CD, Aballea A, Sirigu A. Mapping phantom movement representations in the motor cortex of amputees. Brain. 2006;129(Pt 8):2202–10.
8. Di Pino G, Guglielmelli E, Rossini PM. Neuroplasticity in amputees: main implications on bidirectional interfacing of cybernetic hand prostheses. Prog Neurobiol. 2009;88(2):114–26.
9. Wall JT, Xu J, Wang X. Human brain plasticity: an emerging view of the multiple substrates and mechanisms that cause cortical changes and related sensory dysfunctions after injuries of sensory inputs from the body. Brain Res Brain Res Rev. 2002;39(2–3):181–215.
10. Kaas JH, Florence SL, Jain N. Subcortical contributions to massive cortical reorganizations. Neuron. 1999;22(4):657–60.
11. Pons TP, Garraghty PE, Ommaya AK, Kaas JH, Taub E, Mishkin M. Massive cortical reorganization after sensory deafferentation in adult macaques. Science. 1991;252(5014):1857–60.
12. Ramachandran VS, Stewart M, Rogers-Ramachandran DC. Perceptual correlates of massive cortical reorganization. Neuroreport. 1992;3(7):583–6.
13. Elbert T, Flor H, Birbaumer N, Knecht S, Hampson S, Larbig W, et al. Extensive reorganization of the somatosensory cortex in adult humans after nervous system injury. Neuroreport. 1994;5(18):2593–7.
14. Flor H, Elbert T, Muhlnickel W, Pantev C, Wienbruch C, Taub E. Cortical reorganization and phantom phenomena in congenital and traumatic upper-extremity amputees. Exp Brain Res. 1998;119(2):205–12.
15. Borsook D, Becerra L, Fishman S, Edwards A, Jennings CL, Stojanovic M, et al. Acute plasticity in the human somatosensory cortex following amputation. Neuroreport. 1998;9(6):1013–7.
16. Antfolk C, D'Alonzo M, Controzzi M, Lundborg G, Rosen B, Sebelius F, et al. Artificial redirection of sensation from prosthetic fingers to the phantom hand map on transradial amputees: vibrotactile versus mechanotactile sensory feedback. IEEE Trans Neural Syst Rehabil Eng. 2013;21(1):112–20.
17. Ramachandran VS. Behavioral and magnetoencephalographic correlates of plasticity in the adult human brain. Proc Natl Acad Sci U S A. 1993;90(22):10413–20.
18. Rosen B, Ehrsson HH, Antfolk C, Cipriani C, Sebelius F, Lundborg G. Referral of sensation to an advanced humanoid robotic hand prosthesis. Scand J Plast Reconstr Surg Hand Surg. 2009;43(5):260–6.
19. Merzenich MM, Nelson RJ, Stryker MP, Cynader MS, Schoppmann A, Zook JM. Somatosensory cortical map changes following digit amputation in adult monkeys. J Comp Neurol. 1984;224(4):591–605.
20. Manger PR, Woods TM, Jones EG. Plasticity of the somatosensory cortical map in macaque monkeys after chronic partial amputation of a digit. Proc Biol Sci. 1996;263(1372):933–9.
21. Weiss T, Miltner WH, Huonker R, Friedel R, Schmidt I, Taub E. Rapid functional plasticity of the somatosensory cortex after finger amputation. Exp Brain Res. 2000;134(2):199–203.

22. Lundborg G, Richard P. Bunge memorial lecture. Nerve injury and repair – a challenge to the plastic brain. J Peripher Nerv Syst. 2003;8(4):209–26.

23. Flor H, Elbert T, Knecht S, Wienbruch C, Pantev C, Birbaumer N, et al. Phantom-limb pain as a perceptual correlate of cortical reorganization following arm amputation. Nature. 1995;375(6531):482–4.

24. Knecht S, Henningsen H, Elbert T, Flor H, Hohling C, Pantev C, et al. Cortical reorganization in human amputees and mislocalization of painful stimuli to the phantom limb. Neurosci Lett. 1995;201(3):262–4.

25. Knecht S, Soros P, Gurtler S, Imai T, Ringelstein EB, Henningsen H. Phantom sensations following acute pain. Pain. 1998;77(2):209–13.

26. MacIver K, Lloyd DM, Kelly S, Roberts N, Nurmikko T. Phantom limb pain, cortical reorganization and the therapeutic effect of mental imagery. Brain. 2008;131(Pt 8):2181–91.

27. Schmalzl L, Thomke E, Ragno C, Nilseryd M, Stockselius A, Ehrsson HH. "Pulling telescoped phantoms out of the stump": manipulating the perceived position of phantom limbs using a full-body illusion. Front Hum Neurosci. 2011;5:121.

28. Ramachandran VS, Rogers-Ramachandran D, Cobb S. Touching the phantom limb. Nature. 1995;377(6549):489–90.

29. Rosen B, Lundborg G. Training with a mirror in rehabilitation of the hand. Scand J Plast Reconstr Surg Hand Surg. 2005;39(2):104–8.

30. Schmalzl L, Ragno C, Ehrsson HH. An alternative to traditional mirror therapy: illusory touch can reduce phantom pain when illusory movement does not. Clin J Pain. 2013 Feb 26.

31. Posner MA, Rinaldi E. Upper extremity replantations in Renaissance art. J Hand Surg Am. 2008;33(8):1440–1.

32. Maricevich M, Carlsen B, Mardini S, Moran S. Upper extremity and digital replantation. Hand (N Y). 2011;6(4):356–63.

33. Bjorkman A, Waites A, Rosen B, Lundborg G, Larsson EM. Cortical sensory and motor response in a patient whose hand has been replanted: one-year follow up with functional magnetic resonance imaging. Scand J Plast Reconstr Surg Hand Surg. 2007;41(2):70–6.

34. Muneoka K, Han M, Gardiner DM. Regrowing human limbs. Sci Am. 2008;298(4):56–63.

35. Sanchez Alvarado A. Developmental biology: a cellular view of regeneration. Nature. 2009; 460(7251):39–40.

36. Kumar A, Godwin JW, Gates PB, Garza-Garcia AA, Brockes JP. Molecular basis for the nerve dependence of limb regeneration in an adult vertebrate. Science. 2007;318(5851):772–7.

37. Stappenbeck TS, Miyoshi H. The role of stromal stem cells in tissue regeneration and wound repair. Science. 2009;324(5935):1666–9.

38. Kragl M, Knapp D, Nacu E, Khattak S, Maden M, Epperlein HH, et al. Cells keep a memory of their tissue origin during axolotl limb regeneration. Nature. 2009;460(7251):60–5.

39. Straube WL, Brockes JP, Drechsel DN, Tanaka EM. Plasticity and reprogramming of differentiated cells in amphibian regeneration: partial purification of a serum factor that triggers cell cycle re-entry in differentiated muscle cells. Cloning Stem Cells. 2004;6(4):333–44.

Chapter 15
Hand Transplantation

Abstract Hand transplantation is a controversial surgical procedure that can result in useful hand function but requires lifelong medication with immunosuppressive drugs to prevent rejection. Nerve fibres from the recipient's residual forearm must regenerate into the transplanted hand to reinnervate its intrinsic muscles and sensory receptors. A key issue is restoring sensory functions in the transplanted hand to achieve functional integration. Although more than 70 hand transplants have been performed in more than 50 patients so far, these procedures are only performed in select cases at a few world centres.

Transplantation of kidneys, hearts and livers has improved quality of life for many severely ill people, creating new possibilities for surviving a life-threatening illness. But what about hands? Is it possible to transplant a hand from one individual to another as with a kidney, heart or liver?

The concept of hand transplantation raises several questions. It is easy to assume that such a procedure could be beneficial to people who have lost one or both hands. However, hand transplantation can also create severe problems from the psychological, emotional, ethical, juridical and medical viewpoints. Unlike a heart, kidney or liver transplant to treat a life-threatening illness, many amputees see a hand transplantation as a 'life-giving' procedure, restoring their body image and facilitating their occupational and social integration [1]. Therefore, it is extremely important to weigh the benefits against the risks associated with a hand transplant before making the decision to go ahead with the operation [2–9].

Having a transplanted hand that previously belonged to a deceased individual can create substantial psychological problems: The hand is not hidden inside the body like a transplanted kidney or heart, but is freely exposed to the recipient as well as the environment. A transplanted hand bears all the characteristics of the deceased donor with respect to size, colour, hair growth, birthmarks, nevi, possible scars and even fingerprints. This can be hard for the recipient to accept and adjust to mentally, especially since the transplanted hand will lack sensation and sensory feedback for long time. During this period, the new hand may indeed be perceived as a 'dead man's hand' and feel like a foreign object rather than a true part of the recipient's body.

In the long term, it is possible to achieve fairly good recovery of motor functions as well as some sensory functions in the hand, even if the new hand will never work as well as the recipient's original hand. If the amputee suffers from severe phantom sensations after the amputation, a transplanted hand can 'fill the empty space', helping to ease phantom sensations and phantom pain. Some amputation patients may feel an enormously strong desire to have a 'normal hand' when socialising with others, especially with family members. One patient I met said his greatest wish was to be able to lift and embrace his grandchild and hold him close. Other patients consider it most important to be able to shake hands with others and to use their hands in natural gestures.

From a medical viewpoint, the risks and negative consequences of a hand transplant may be quite substantial. As with other types of organ transplants, the body's immunological system perceives the hand transplant as a foreign object that must be rejected. The consequence is an acute strong immunological reaction when the body tries to reject the new hand [8, 10–12]. A hand consists of several kinds of tissues with strong antigenic properties that rapidly activate the recipient's immunological system. The skin has especially strong antigenic properties. Over the years, the experiences from transplanting internal organs have been helpful in developing successively better and more effective immunosuppressive drugs. Various types of steroids still have an important role, and cyclosporine, discovered in the 1980s, proved to be an effective immunosuppressant. Tacrolimus, which came in to use about 20 years ago, also proved very effective in inhibiting rejection, and since then several other substances with various modes of action have been developed. The special and complex immunological problems associated with hand transplantation also apply to face transplants, since both are complex tissue allografts (CTAs) characterised by multiple tissues with high antigenic properties. Today the traditional approach to the immunosuppressive regimen is initial treatment with induction agents such as polyclonal (antithymocyte globulins) or monoclonal (alemtuzumab, basiliximab) antibody preparations followed by a maintenance therapy including tacrolimus (FK-506), mycophenolate mofetil (MMF) and prednisone [13, 14]. The immunosuppressive medications must be continued throughout the patient's lifetime and are associated with several serious risks and side effects, including a severely impaired defence against infections leading to a risk of sepsis, risk for developing diabetes and malignant tumours as well as cardiovascular problems. An amputee wanting to have a hand transplant must be aware of all these risks and ask himself if the benefits of the procedure are really worth all the negative effects [5, 15].

From the surgical point of view, a hand transplant is not very complicated; the procedure can be performed under well-controlled conditions. The potential problems and risks from the transplantation itself and the subsequent medications make it a controversial procedure, and over the years hand transplantation has only been performed with certain restrictions on a limited number of patients and at a limited number of the world's replantation centres [6].

Hand Transplantation from an International Perspective

The first hand transplantation was performed in 1964 in Ecuador by Dr Robert Gilbert [16, 17]. The hand had to be amputated after 3 weeks due to acute rejection; the surgeons were not sufficiently knowledgeable about the medication required to inhibit rejection. A 30-year period of stagnation followed.

The second hand transplantation was performed on 25 September 1998 at the Edouard Herriot Hospital in Lyon, France, by a team led by surgeons Jean-Michel Dubernard, Earl Owen from Sydney and Nadey Hakim from London [18]. A hand from a recent motorcycle accident victim was transplanted onto 48-year-old New Zealander Clint Hallem, who was brought to Lyon for the procedure. The operation took 14 h and was hailed as a breakthrough in the world press. Unfortunately, it ended in disappointment as it became necessary to amputate the hand after 3 years. The patient, Clint Hallem, appeared on TV and blamed himself. He also told the *New York Times* how he had not been able to continue taking the heavy medication and how, on his own initiative, he had stopped the immunosuppressive medications necessary to inhibit rejection. Hallem also made clear that he had not been able to accept the new body part and he felt 'mentally isolated' from the hand. This was a severe disappointment, and it was now realised that selecting suitable patients is essential for a successful hand transplant.

On 25 January 1999, just a few months after the hand transplantation in Lyon, the first American hand transplantation was performed in Louisville, Kentucky, by a team led by Warren Breidenback [2]. The patient, Matthew Scott, had lost a hand in a fireworks accident several years earlier. The surgery went well and Scott is today the longest surviving successful hand transplant recipient in the world.

The first double-hand transplantation was performed in January 2000 in Lyon when the 33-year-old Denis Chatelier received two new hands from a deceased man. Chatelier had lost his hands when a homemade rocket exploded. Soon thereafter, on 17 March of the same year, a double-hand transplant was performed at University Hospital Innsbruck by Raimund Margreiter and colleagues [4, 19]. After about 6 months this patient began to recover sensation in the hands, and after 1 year he could shave himself, hold a pencil and use a pair of scissors [7]. In the ensuing years, several successful single- and double-hand transplantations were performed in several centres in the USA, Europe, South America and Asia [14]. The first complete double arm transplantation was performed in August 2008 by Christoph Hoehnke and Edgar Biemer on farmer Karl Merck, who had lost both arms in a 2002 combine accident [20]. The surgery was performed in Munich in the hospital that Merck himself had approached after watching a TV programme about hand transplants and after trying to make several homemade arm prostheses. The surgery was successful and without complications.

A few months after the surgery, Merck could move his new fingers and hands since the tendons and muscles in the hand transplants were attached to corresponding muscles in the recipient's remaining arm stumps. After intense physiotherapy, he rapidly gained the ability to open doors and operate a light switch.

Although most hand transplants are performed to replace hands that are lost due to traumatic amputation, the indications may vary. In October 2010, Dr Marco Lanzetta at the Italian Institute of Hand Surgery in Monza (Milan in Italy) performed a bilateral hand transplant on a 52-year-old female who suffered quadrimembral amputation as a result of sepsis. Britain's first hand transplant was carried out on 27 December 2011 by a team led by Dr Simon Kay at Leeds General Infirmary. Mark Cahill, a 51-year-old former publican, had been left with a non-functioning right hand as a result of gout and a subsequent infection. In a newspaper interview in the *Telegraph* on 4 January 2013, Cahill claimed that he had already gained some movement in his fingers following the operation. He expressed delight at his 'brand-new hand', hoping that it would enable him to once again cut his own food, dress himself and play properly with his grandson.

All hand transplants that are performed worldwide are registered and closely monitored in the *International Registry on Hand and Composite Tissue Transplantation* (www.handregistry.com) [6, 20]. To date more than 70 hand transplants have been performed on more than 50 patients in various parts of the world (www.handregistry.com, accessed 10 April 2013). This means that more than 20 patients have received two new hands at the same time.

Since the first transplant on Clint Hallem in 1998, emphasis has been placed on the risk of acute rejection episodes. Often there are episodes of beginning rejection that need immediate medical treatment. A hand consists of several types of tissues with varying antigenicity, and the tendency for rejection can vary. The early stages of a rejection reaction are first obvious in the skin, so it is advantageous that the hand is fully visible and not hidden in the body as in other types of organ transplantations [21].

Transplanted hands can have fairly good function, at least as good as if the patient's own hand had been *replanted* after an amputation [8, 18, 22–25]. Muscles and tendons in the transplanted hand are attached to the recipient's corresponding muscles and tendons. Consequently, the recipient may be able to move the fingers in the transplanted hand after only a few days. Recovery of sensation in the transplanted hand is a much more complicated issue but necessary to achieve functional integration of the new hand. The recipient's nerve trunks, measuring 2–3 mm in diameter, are adapted to the corresponding nerve trunks in the transplanted hand. Regenerating nerve fibres from the recipient's nerve trunks then have to advance and grow down the endoneural tubes, i.e. the Schwann cell tubes in the nerve trunks of the transplanted hand, to reinnervate its peripheral sensory receptors – a time-consuming process requiring months and years before there is any recovery of useful sensibility in the transplanted hand.

The immunosuppressive treatment has no negative effect on nerve regeneration. On the contrary, it has been argued by some scientists that treatment with tacrolimus (FK-506) may even promote nerve regeneration [26–28]. Another problem is the misdirection of regenerating nerve fibres that always occurs to a great extent, which results in a remapping of the cortical hand representation into a mosaic-like pattern [29]. Nevertheless, in several cases transplanted hands may recover a fairly useful sensibility with the ability to feel temperature, pain and pressure stimuli applied to the transplanted hand. The International Registry on Hand and Composite

Transplantation, available at www.handregistry.com, reviewed hand transplants performed over an 11-year period (September 1998–July 2010) and reported on outcomes [14, 30]. It was reported that all patients regained protective sensibility and that 82 % developed discriminative sensibility in the transplanted hand. Recovery of muscle function in the hand allowed patients to perform most activities of daily living. Moreover, 75 % of the recipients reported an improvement in quality of life, and many had returned to work.

The first two American hand transplant patients have been followed for more than 11 and 9 years, respectively. In a follow-up after 8 and 6 years, respectively, both patients were back at their own occupations, and they enjoyed a very good quality of life. They had good function in their hands and could perform several types of activities of daily living such as tying shoelaces, turning pages in a newspaper, writing with a pencil and picking up small items [22].

The use of brain imaging techniques, such as functional magnetic resonance imaging (fMRI), demonstrated that over time a transplanted hand 'regains' its normal position in the recipient's brain and consequently will be perceived as integrated in and belonging to the body [31–35].

A Controversial Procedure

Today hand transplantation is regarded by many as controversial and questionable. It has often been argued that the extensive and stressful lifelong immunosuppressive medication can only be justified in association with life-saving surgical procedures [36]. But if the amputee strongly desires a new hand with motor and sensory functions, many patients may be prepared to accept the risks that are associated with the surgical procedure and the required medication. Those who have lost only one hand can usually manage well by using the remaining healthy hand, but the situation can be completely different for those who have lost both hands. It is a common opinion among many surgeons that hand transplantation is justified only if both hands are amputated.

No hand transplantation has so far been performed in Sweden, but I have personally been involved in two hand transplantations in Monza, Italy, in cooperation with Professor Marco Lanzetta. One of the biggest postsurgical problems after hand transplantation is that for a long time, maybe half a year, the patient totally lacks sensibility in the hand. The role of my colleagues and myself during this initial period was to provide the patients with a kind of 'artificial' sensibility based on a so-called Sensor Glove equipped with miniature microphones at finger level (Chap. 10). The concept is that the friction sounds that are generated when the hand touches items and textures are transferred to two earphones so that the patient can listen to what the hand touches. It was previously demonstrated that activation of the cortical hearing area in such a situation also generates an activation of the nearby somatosensory cortex via so-called audio-tactile interaction [37]. Using this principle the recipients' transplanted hands were equipped with a Sensor Glove from the first day after surgery to achieve sensory feedback. Early use of this technique helped the patient perceive the new hand as being integrated with the body, and in fMRI studies the transplanted hand regained a correct position in the recipients' somatosensory cortices [35].

One patient argued that his quality of life was so greatly improved by the transplant that he did not feel the slightest hesitation about the value of the surgery – 'the hand transplantation was perhaps not lifesaving but it was definitely life-giving'. I could easily agree with him when, 6 months after the transplantation, I could see with my own eyes how he socialised and shook hands with people around him using the transplanted hand, telling them about how he was now able to play ball games with his grandchildren.

John Irving described with great clever humour and insight the problem complex associated with hand transplants in his novel *The Fourth Hand* [38]. In it, a well-known journalist, Patrick Wallingford, conducting a live online feature on a circus in India, got too close to the lion cage and before the eyes of millions of viewers his left hand was caught and eaten by a starving lion. A woman in Wisconsin, Doris Clausen, was willing to donate the left hand of her deceased husband who had died of a self-inflicted gunshot wound, however, with the unorthodox request that she has hand visitation rights.

The hand was successfully attached to Patrick Wallingford's forearm by Dr Nicholas M. Zajac of Boston, who had long been awaiting the opportunity to perform the nation's first hand transplant. However, the situation was complicated by the fact that Patrick fell in love with Doris. She, however, only allowed him to touch her intimately with her husband's hand, now residing on Patrick's forearm.

The Future

More than anything else, it is the lifelong and risky immunosuppressive medication that makes hand transplantations controversial. However, considering the development and trends in immunological research, the situation may be different in the future. New possibilities are evolving based on a new principle: Rather than suppressing rejection through pharmacological intervention, the idea is to develop *tolerance* for the hand transplant in the recipient without requiring heavy medication. Laboratory studies have shown it is possible to develop a tolerance for a transplanted organ by transplanting cells from the donor's bone marrow to the recipient after first having knocked out the recipient's bone marrow with high doses of radiation or by some other means. This allows the creation of a cell population that possesses the donor's as well as the recipient's immunological properties, meaning that the recipient's biological system does not perceive the transplanted tissue as foreign [10, 39–42]. By creating this *chimerism*, the goal is to induce donor-specific tolerance. Tolerance may be defined as hyporesponsiveness to the donor without immunosuppression while maintaining adequate immune response to combat third-party antigens [14, 43]. Successful initial clinical trials to combine clinical transplantation with infusion of donor bone marrow are presently being performed at some centres

[13, 44, 45]. If this principle 1 day gains a broader clinical application, it could mean a revolution in all transplantation activities, especially hand transplants. It would make it possible to transplant hands or parts of hands without the additional severe and risky immunosuppressive medications.

References

1. Lanzetta M, Dubernard JM, Petruzzo P. Hand transplantation. Milan: Springer; 2007.
2. Breidenbach 3rd WC, Tobin 2nd GR, Gorantla VS, Gonzalez RN, Granger DK. A position statement in support of hand transplantation. J Hand Surg Am. 2002;27(5):760–70.
3. Dubernard JM, Henry P, Parmentier H, Vallet B, Vial D, Badet L, et al. First transplantation of two hands: results after 18 months. Ann Chir. 2002;127(1):19–25.
4. Margreiter R, Brandacher G, Ninkovic M, Steurer W, Kreczy A, Schneeberger S. A double-hand transplant can be worth the effort! Transplantation. 2002;74(1):85–90.
5. Lanzetta M, Petruzzo P, Vitale G, Lucchina S, Owen ER, Dubernard JM, et al. Human hand transplantation: what have we learned? Transplant Proc. 2004;36(3):664–8.
6. Lanzetta M, Petruzzo P, Margreiter R, Dubernard JM, Schuind F, Breidenbach W, et al. The International Registry on Hand and Composite Tissue Transplantation. Transplantation. 2005; 79(9):1210–4.
7. Schneeberger S, Ninkovic M, Piza-Katzer H, Gabl M, Hussl H, Rieger M, et al. Status 5 years after bilateral hand transplantation. Am J Transplant. 2006;6(4):834–41.
8. Schuind F, Abramowicz D, Schneeberger S. Hand transplantation: the state-of-the-art. J Hand Surg Eur Vol. 2007;32(1):2–17.
9. Cooney WP, Hentz VR. Hand transplantation – primum non nocere. J Hand Surg Am. 2002; 27(1):165–8.
10. Siemionow M, Ortak T, Izycki D, Oke R, Cunningham B, Prajapati R, et al. Induction of tolerance in composite-tissue allografts. Transplantation. 2002;74(9):1211–7.
11. Siemionow M, Ozer K. Advances in composite tissue allograft transplantation as related to the hand and upper extremity. J Hand Surg Am. 2002;27(4):565–80.
12. Brandacher G, Lee WP. Hand transplantation. Hand Clin. 2011;27(4):xiii–xiv.
13. Ravindra KV, Ildstad ST. Immunosuppressive protocols and immunological challenges related to hand transplantation. Hand Clin. 2011;27(4):467–79, ix.
14. Foroohar A, Elliott RM, Kim TW, Breidenbach W, Shaked A, Levin LS. The history and evolution of hand transplantation. Hand Clin. 2011;27(4):405–9, vii.
15. Dickenson D, Widdershoven G. Ethical issues in limb transplants. Bioethics. 2001;15(2):110–24.
16. Barker JH, Francois CG, Frank JM, Maldonado C. Composite tissue allotransplantation. Transplantation. 2002;73(5):832–5.
17. Gilbert R. Transplant is successful with a cadaver forearm. Med Trib Med News. 1964;5:20–3.
18. Dubernard JM, Owen E, Herzberg G, Lanzetta M, Martin X, Kapila H, et al. Human hand allograft: report on first 6 months. Lancet. 1999;353(9161):1315–20.
19. Hautz T, Engelhardt TO, Weissenbacher A, Kumnig M, Zelger B, Rieger M, et al. World experience after more than a decade of clinical hand transplantation: update on the Innsbruck program. Hand Clin. 2011;27(4):423–31, viii.
20. International Registry on Hand and Composite Tissue Transplantation. Available at: www.handregistry.com. Accessed Apr 2013.
21. Cendales LC, Kirk AD, Moresi JM, Ruiz P, Kleiner DE. Composite tissue allotransplantation: classification of clinical acute skin rejection. Transplantation. 2005;80(12):1676–80.
22. Breidenbach WC, Gonzales NR, Kaufman CL, Klapheke M, Tobin GR, Gorantla VS. Outcomes of the first 2 American hand transplants at 8 and 6 years posttransplant. J Hand Surg Am. 2008;33(7):1039–47.

23. Dubernard JM, Petruzzo P, Lanzetta M, Parmentier H, Martin X, Dawahra M, et al. Functional results of the first human double-hand transplantation. Ann Surg. 2003;238(1):128–36.

24. Francois CG, Breidenbach WC, Maldonado C, Kakoulidis TP, Hodges A, Dubernard JM, et al. Hand transplantation: comparisons and observations of the first four clinical cases. Microsurgery. 2000;20(8):360–71.

25. Herzberg G, Parmentier H, Erhard L. Assessment of functional outcome in hand transplantation patients. Hand Clin. 2003;19(3):505–9, x.

26. Fansa H, Keilhoff G, Altmann S, Plogmeier K, Wolf G, Schneider W. The effect of the immunosuppressant FK 506 on peripheral nerve regeneration following nerve grafting. J Hand Surg Br. 1999;24(1):38–42.

27. Kvist M, Danielsen N, Dahlin LB. Effects of FK506 on regeneration and macrophages in injured rat sciatic nerve. J Peripher Nerv Syst. 2003;8(4):251–9.

28. Glaus SW, Johnson PJ, Mackinnon SE. Clinical strategies to enhance nerve regeneration in composite tissue allotransplantation. Hand Clin. 2011;27(4):495–509, ix.

29. Lundborg G. Nerve injury and repair. Regeneration, reconstruction and cortical remodelling. 2nd ed. Philadelphia: Elsevier; 2004.

30. Petruzzo P, Lanzetta M, Dubernard JM, Landin L, Cavadas P, Margreiter R, et al. The International Registry on Hand and Composite Tissue Transplantation. Transplantation. 2010;90(12):1590–4.

31. Cavadas PC, Landin L, Ibanez J. Bilateral hand transplantation: result at 20 months. J Hand Surg Eur Vol. 2009;34(4):434–43.

32. Frey SH, Bogdanov S, Smith JC, Watrous S, Breidenbach WC. Chronically deafferented sensory cortex recovers a grossly typical organization after allogenic hand transplantation. Curr Biol. 2008;18(19):1530–4.

33. Brenneis C, Loscher WN, Egger KE, Benke T, Schocke M, Gabl MF, et al. Cortical motor activation patterns following hand transplantation and replantation. J Hand Surg Br. 2005;30(5):530–3.

34. Roricht S, Machetanz J, Irlbacher K, Niehaus L, Biemer E, Meyer BU. Reorganization of human motor cortex after hand replantation. Ann Neurol. 2001;50(2):240–9.

35. Lanzetta M, Perani D, Anchisi D, Rosen B, Danna M, Scifo P, et al. Early use of artificial sensibility in hand transplantation. Scand J Plast Reconstr Surg Hand Surg. 2004;38(2):106–11.

36. Lanzetta M, Nolli R, Borgonovo A, Owen ER, Dubernard JM, Kapila H, et al. Hand transplantation: ethics, immunosuppression and indications. J Hand Surg Br. 2001;26(6):511–6.

37. Lundborg G, Bjorkman A, Hansson T, Nylander L, Nyman T, Rosen B. Artificial sensibility of the hand based on cortical audiotactile interaction: a study using functional magnetic resonance imaging. Scand J Plast Reconstr Surg Hand Surg. 2005;39(6):370–2.

38. Irving J. The fourth hand. London: Random House; 2001.

39. Kanitakis J, Jullien D, Claudy A, Revillard JP, Dubernard JM. Microchimerism in a human hand allograft. Lancet. 1999;354(9192):1820–1.

40. Hettiaratchy S. Transplantation tolerance and chimerism: what are they and do we need them? Plast Reconstr Surg. 2004;113(7):2213–4.

41. Hettiaratchy S, Melendy E, Randolph MA, Coburn RC, Neville Jr DM, Sachs DH, et al. Tolerance to composite tissue allografts across a major histocompatibility barrier in miniature swine. Transplantation. 2004;77(4):514–21.

42. Siemionow M, Klimczak A. Tolerance and future directions for composite tissue allograft transplants: part II. Plast Reconstr Surg. 2009;123(1):7e–17.

43. Wu S, Xu H, Ravindra K, Ildstad ST. Composite tissue allotransplantation: past, present and future-the history and expanding applications of CTA as a new frontier in transplantation. Transplant Proc. 2009;41(2):463–5.

44. Kawai T, Cosimi AB, Spitzer TR, Tolkoff-Rubin N, Suthanthiran M, Saidman SL, et al. HLA-mismatched renal transplantation without maintenance immunosuppression. N Engl J Med. 2008;358(4):353–61.

45. Scandling JD, Busque S, Dejbakhsh-Jones S, Benike C, Millan MT, Shizuru JA, et al. Tolerance and chimerism after renal and hematopoietic-cell transplantation. N Engl J Med. 2008;358(4):362–8.

Chapter 16
The Mind-Controlled Robotic Hand

Abstract A mind-controlled artificial hand is a good solution for improving quality of life for many amputees. However, considering the extremely well-developed motor and sensory functions of the human hand, developing a useful alternative is an enormously challenging task. The ideal hand prosthesis should be capable of intuitively executing the delicate movements and precision grips that we use daily at work and at leisure. It should possess sensory functions to ensure a feeling of embodiment, and it should provide sensory feedback for regulation of grip strength. It should also provide tactile discriminative functions. However, despite advanced technological achievements in the prosthetic field, only some of these aspirations have been fulfilled. The prosthetic hands that are available on the market today are controlled by EMG signals from the forearm muscles, and their function is limited to little more than opening and closing the hand. Various principles for providing them with sensory functions are currently being tried in many centres, and refined motor functions are being achieved by using computerised systems for pattern recognition, associating specific patterns of myoelectric signals with specific movements of the hand. In experimental studies and clinical trials on completely paralysed patients, recordings from needle electrodes, implanted into motor areas of the brain, have been successfully used to control movements in artificial arms and hands. In the future the function of mind-controlled robotic hands will presumably come much closer to the function of a human hand.

Substituting an artificial hand for a missing hand has long been a well-known concept. Cosmetic prostheses with the appearance of normal hands were used by the Egyptians thousands of years ago. Various types of hand prostheses were constructed during the sixteenth and seventeenth centuries that were able to open and close with help from the healthy hand, and various types of spring mechanisms were used to give a firm grip, for instance, around the hilt of a sword. There is a story about a German knight called Götz von Berlichingen who, at the age of 24, lost his right hand in battle. Götz constructed an iron hand with movable fingers so

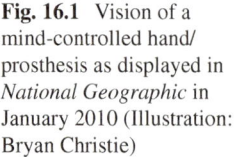

Fig. 16.1 Vision of a mind-controlled hand/prosthesis as displayed in *National Geographic* in January 2010 (Illustration: Bryan Christie)

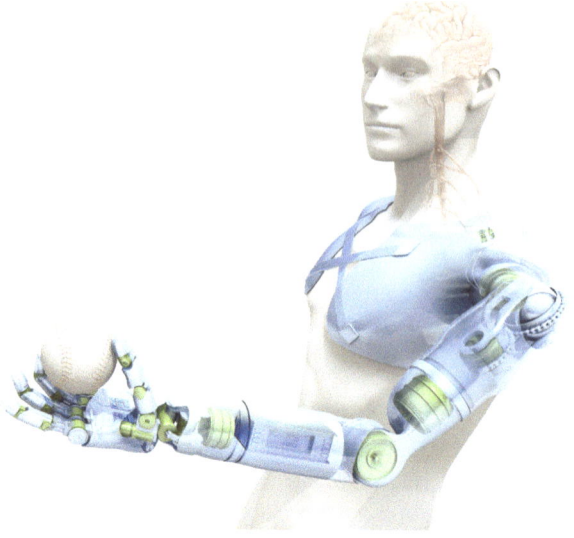

he could grip a sword and a lance. With his 1.4 kilo heavy iron hand, he continued to actively serve on the battlefield for 40 more years. He felt that the iron hand worked at least as well in battle as his real hand. Today Götz's iron hand is on display at a museum in Nuremberg.

In recent times, interest in developing hand and arm prostheses has been quite limited, and for a long time cosmetic prostheses, or simple hooks controlled by shoulder movements, were the only prostheses available. The number of people with amputations at the arm and forearm levels has been limited in comparison to patients with lower limb amputations, as diabetes often makes it necessary to amputate a leg below knee level. However, the thalidomide disaster in the late 1960s and early1970s created an evolving interest in constructing artificial hands [1–5].

A functional artificial hand for amputees must fulfil certain requirements. It should be based on a mechanical construction that allows movements and grips that are useful for the amputee, and the hand's performance should be intuitively controlled by the user's mind. The user's thoughts and intentions should be automatically transformed into movements of the prosthesis. This vision represents an enormous biomechanical and neurobiological challenge (Fig. 16.1) [6].

In movies and computer games, it is easy to be impressed by how easily advanced robotic hands can mimic the function of normal hands and how easily they can be repaired or replaced if needed. But in reality we have a long way to go before reaching such a goal. Making a mind-controlled hand move and function intuitively based on our thoughts requires some kind of functional connection between the prostheses and the nervous system – a man–machine interface. In theory there are various possibilities for achieving this goal. A limited functional connection can be established with the brain by recording brain activity using electrodes on the outside of the skull, like in EEG examinations. However, a more precise and accurate recording of brain activity requires the use of thin needle electrodes implanted directly into the brain. This technique has frequently been used in animal experiments, and such a direct connection between the brain and a hand prosthesis, using advanced control algorithms for pattern recognition, has also been used in a small number of totally paralysed patients.

A simpler alternative to establish contact with the nervous system would be to place recording electrodes around or inside a nerve trunk in the residual arm of the amputee and to use the electrical activity in the nerve to control movements in the hand prosthesis via cables penetrating the skin. However, this technique is associated with several biological and mechanical problems. A far more attractive concept, which today is routinely used in all commercially available mind-controlled prostheses, is to record and utilise the electrical signals generated by muscle activity in the residual forearm, so-called EMG (electromyographic) signals. Even if the hand is amputated, it is still possible to imagine movements in the missing hand [7]; the motor area of the brain is activated, and electrical signals are transmitted to those muscles in the forearm that would normally regulate hand movements.

Controlling an Artificial Hand Through Direct Recordings from the Brain

The principle of controlling an artificial hand using recordings directly from the brain was clinically tested in order to create communication with the outside world for totally paralysed patients with 'locked-in syndrome', a total isolation from the outer world because of a spinal injury close to the brain or a severe neurological disease [8–19]. In 2000, there were reports of three patients with 'locked-in syndrome' where fine needle electrodes were implanted in the brain to record the electrical activity in specific groups of motor nerve cells [10]. The subjects were able to voluntarily raise and lower a cursor sweeping horizontally across a computer screen, thus making the cursor hit a vertical row of letters on the right edge of the screen. After training, the paralysed patients were able to hit specific letters and spell out words. Among the first words that one of the patients spelled were the names of two of the scientists who had made it possible for him to communicate with the outside world for the first time, the sequence of letters being 'KENEDY GQLDXWAIJTEF' (Kennedy, Goldwaithe) [10].

From the beginning, the objective in these trials was to record signals from individual nerve cells, but it was soon realised that an array containing a large number of microelectrodes gave the best results [8, 19–21]. Miguel Nicolelis at Duke University in Durham, North Carolina, is a pioneer when it comes to transforming thoughts into activities and movements in artificial extremities, using microelectrode arrays introduced into the brain [20, 21]. His early laboratory experiments involved implanting microelectrodes into the motor cortices of monkeys [22–25]. The signal patterns that were generated when the monkey performed simple trained movements in the hand and arm were recorded by the electrode arrays and were processed in an artificial neural network, which learned to associate specific signal patterns with specific arm movements. The system was connected to a robot arm that could then perform the corresponding movements simultaneously with the monkey's own movements: as soon as the monkey initiated an arm movement, the associated signal pattern had already been registered and interpreted by the computer, and an identical movement was simultaneously performed by the robot arm.

The monkey, Belle, became a pioneer in these trials when, one afternoon in the spring of 2000, she was able to voluntarily control movements not only in a robotic arm positioned in the room beside her but also in another robotic arm in a laboratory more than 100 miles away via wireless transfer [24]. The experiments were performed at Duke University, but by linking the computers together with the Laboratory for Human and Machine Haptics in Cambridge, Massachusetts, the thoughts generated in the monkey's brain were able to induce identical and simultaneous movements in the robotic arms in both places.

It has even proved possible to bypass the monkey's thoughts and to 'read' the activities already generated in the brain's 'planning centre' in the premotor cortex, where decisions are made to perform specific movements. In this way the robotic arm performed the movements before the movements were actually performed by the monkey's arm [26]. In 2008 Schwartz and colleagues in Pittsburgh reported a new generation of trials where, after needle electrodes were implanted in their motor cortices, monkeys could be trained to perform advanced natural movements in a robotic arm in three dimensions [27].

The same principle has been applied on a small scale in patients with 'locked-in syndrome'. In the *BrainGate Project*, Hochberg and colleagues in Massachusetts and Boston introduced a 4×4 mm array of close to 100 microelectrodes in the motor cortex of a human with a long-standing tetraplegia subsequent to a brain stem stroke, unable to move any muscles in his upper or lower extremities. The signal patterns that were generated by the patient's thoughts were decoded in advanced pattern-recognition systems and could be translated into several various activities like opening email, playing computer games and even initiating movements in the hand prosthesis, functions that were well preserved after follow-up 1,000 days after implantation [28, 29]. In an analogous case Schwartz, Collinger and colleagues in Pittsburgh, Pennsylvania, recently found that the participant was able to control and move the prosthetic limb in the three-dimensional workspace on the second day of training and after 13 weeks was able to perform multidimensional movements [30]. The interpretation of these observations was that the implantation of microelectrodes in the motor cortex is a promising principle to improve the quality of life for paralysed patients.

For several years, a developmental project along the same line has been ongoing at the Neuronano Research Center at Lund University in Sweden, with the goal of developing a new generation of extremely thin electrodes using nanotechnology, for permanent implantation in the brain or the spinal cord [31–34]. These electrodes need to be made of a material that is accepted by the brain even in a long-term perspective without inducing any harmful tissue reactions. Still the problem remains how to create a functional brain-machine interface to transmit the recordings from the brain through the skull bone to the computer equipment on the outside of the skull so that the signals can be interpreted and utilised to regulate movements in an artificial hand or handle a computer, a wheelchair or robotic devices.

If the electrodes can be permanently implanted into the brain, it could open a dramatic new landscape for the future for rehabilitation of severely disabled individuals and introduce new types of ethical problems that we have not previously encountered. When a man–machine interface is created, where does the individual end and the machine begin? Who is responsible for involuntary actions that may be performed by a machine or a robotic device linked to the nervous system of a patient via a computerised man–machine interface? And who takes responsibility for the long-term risks that are associated with long-term implantation of needle electrodes in the brain? There are many questions, but so far few answers.

With electrodes in the brain, our wishes and thoughts can be translated into controlled processes in computers, robotic devices and artificial hands. But we are dealing with a principle that makes transmission of signals in both directions possible, even into the brain. Externally generated electrical signals can be used to stimulate specific areas of the brain, a phenomenon that is already used in the treatment of Parkinson's disease (deep brain stimulation) and chronic pain.

However, the possibility of linking the human brain to a computer also creates a frightening outlook for the future [35]. Although in the long term it may be possible to treat specific psychiatric disorders, it will also be possible to eliminate empathic and inhibiting systems to produce, for instance, effective soldiers. But perhaps it will 1 day be possible to boost fading memory functions or to improve learning and other intellectual functions or to plug in a USB stick containing a dictionary program to make a vacation trip to a foreign country more enjoyable.

Controlling an Artificial Hand Using Electrodes Implanted in a Nerve

Normally hand movements are based on the activation of the motor cortex and electrical signals being transmitted to the hand and forearm via the spinal cord and nerve trunks in the arm. Even after a hand amputation, it is still possible to imagine movements in the hand that no longer exists, thereby activating the cortical motor areas [7]. One attractive concept is to utilise direct recordings of electrical activity in the nerves to control the hand prosthesis. Clinical trials in that direction have been carried out [36–38]. However, this technique is associated with several problems regarding obtaining recordings from appropriate fibre components in the nerve and managing the issue of skin-penetrating wires.

Kevin Warwick at the University of Reading in the UK performed some spectacular clinical experiments in 2003 to demonstrate the potential of direct recordings from peripheral nerves [39]. These trials had great media impact. Warwick underwent a surgical procedure on his own arm, introducing an array containing hundreds of microelectrodes into his own median nerve at the distal forearm level.

The median nerve is the major sensory nerve in the hand, which also innervates several of the thumb muscles, so the procedure was far from risk-free.

When Warwick moved his hand, the electrodes recorded the nerve signals generated in the median nerve. The microelectrode array was connected to a computer via skin-penetrating wires, and by using this equipment Warwick was able to control movements in a hand prosthesis and even to control movements of an electrical wheelchair by activating his median nerve.

Intuitively Controlling an Artificial Hand Through EMG Signals: Myoelectric Prostheses

An attractive concept for controlling movements in a mind-controlled hand prosthesis is to utilise the EMG signals elicited by muscle activities in the residual forearm [2, 3]: the hand prostheses currently available on the market, so-called myoelectric prostheses, are all based on this principle. The motor signals from the brain that reach the muscles of the residual forearm have been sorted out along the way to various groups of muscles that normally interact to perform intuitive specific hand movements and grips. Surface electrodes positioned on the skin on top of the extensor and flexor muscles can monitor the contractions and the electrical activity in these muscles, and the generated EMG signals can be used to open and close a prosthetic hand. The principle is simple and ingenious, and this type of prosthesis has been available on the market for several decades. But these prostheses have a very simple function and are often perceived by the amputees as primitive and slow in their actions.

Dr Rolf Soerbye, head of the Department of Neurophysiology in the city of Örebro, Sweden, was a pioneer in the development of a myoelectric hand prosthesis for children. Soerbye used EMG signals – electrical signals generated by contractions in the forearm muscles – to control movements in the hand prosthesis [1]. The simple mechanical construction of the hands allowed for voluntary opening and closing of the prosthetic hand. Soerbye used an effective trick to make the training enjoyable for the children – he let them control small electric toy cars using EMG signals from their forearm. The children were able to make the cars move in various directions by activating the extensor or flexor muscles in their forearms.

Conventional myoelectric prostheses usually use surface electrodes. One electrode, placed on the volar aspect of the forearm, 'listens to' and records contractions and electrical activity in the flexor muscles, while the other electrode is positioned dorsally on top of the extensor muscles to record extensor muscle activity. Voluntary activation of the flexor muscles activates an electric motor in the hand prosthesis, allowing it to close, while activating the extensor muscles opens the hand. However, by using more than two recording electrodes, the system can define *patterns* of muscle signals and associate them with specific movements using pattern-recognition algorithms, with the aim of achieving more complex movements in the artificial hand. This principle was put to use in the early 1990s by a research team at

Sahlgrenska Hospital at Gothenburg University, to develop a mind-controlled hand prosthesis, called the SVEN hand. But the hand was too fragile, and the technology was not sufficiently advanced to make long-term daily use possible [40].

After the first clinical trials, further development of hand prostheses was at a standstill for a long time. Although the available myoelectric prostheses can often be very useful, many hand amputees refuse to use them [41, 42]. They often feel that the prostheses are too primitive and too slow in their movements [4, 5]. However, a major reason for the limited use of hand prostheses is the lack of sensibility; there is no sensory feedback from the artificial hand, and as a consequence they are not perceived as a natural part of the body – the brain does not develop a sense of body ownership of the prosthesis [43].

With the evolution of more advanced technology, recent decades have seen prostheses become more user-friendly [6]. In addition, there is an increasing general interest in the development of limb prostheses since the wars in Iraq, Afghanistan and the Middle East have resulted in an increasing number of amputation injuries. In the USA, Canada and the UK, a substantial amount of money is being invested in developing technically advanced hand and arm prostheses, and there are also several ongoing projects within the European Union. The goal is to develop a new generation of mind-controlled artificial hands with improved electronics and new materials capable of performing most of the functions of a normal hand, for instance, gripping a small ball or pencil, picking up and handling a key, pouring water from a bottle, lifting a glass of water and picking up small items from a flat surface. But one determining factor for how well the hand prosthesis can be perceived as belonging to the body is the sensory feedback it gives. Achieving sensory feedback in a prosthetic hand is an enormous challenge that is the subject of several current research projects in the field of prosthetics.

For many years, developmental projects aimed at such 'intelligent' myoelectric prostheses have been going on at several of the world's research centres and laboratories [44–61]. In many of these projects, the concept is to incorporate several miniature electric motors into the hand mechanics aiming at individual mobility in the thumb and fingers. An important part of the construction is that the hand is 'adaptive' and automatically adjusts its grip to the shape of the item in question. When such an artificial hand grips a mug, the thumb, index and middle fingers close round the mug first, while the ring and little fingers continue the closing action until they too encircle the mug (Fig. 16.2).

The Austrian company Otto Bock has long dominated the hand prosthetic market and is continuously improving its models. Early models were based on one isolated motor that could open and close the prosthetic hand, but in new models, such as the Michelangelo hand, there are two separate motor units for the fingers and thumb, respectively, making more grip functions possible. The most recent model allows the wrist to be passively placed in several positions. There is currently ongoing research to create motorised wrists capable of rotation and flexion.

Several international companies currently have hopes of developing improved, more advanced hand prostheses. Among several models on the market that allow for various grip functions is the i-Limb prosthesis, developed in Edinburgh and

Fig. 16.2 The artificial hand developed in the SmartHand project grips a cup. The hand adapts the grip to the shape of the mug to make the grip effective (Image supplied by Marco Controzzi)

manufactured by Touch Bionics. The i-Limb Hand is made in modules with one separate motor incorporated in each finger, making it simple to replace separate parts of the hand if mechanical problems should occur. The construction allows several grip positions, such as handling a key or picking up small items from a flat surface. Other models with multiple incorporated motors and many useful grip functions are currently under development by companies in the USA, the UK, Germany, Italy and China.

Sweden and Italy are involved in a joint project to develop a new generation of mind-controlled hand prostheses with numerous grip functions combined with sensory functions. In the EU project SmartHand, a mechanically advanced hand based on four motors was constructed at *the BioRobotics Institute*, Scuola Superiore, Sant'Anna, Pisa, Italy [45, 55, 58, 59, 62] (Fig. 16.2), and a principle for refined myoelectric control of the prosthesis was developed using pattern-recognition algorithms [45, 62–64]. Specific training protocols were developed to achieve intuitive control of the prosthesis based on the principle that specific EMG patterns recorded from the amputation stump are associated with specific hand movements and grip functions. The result is a prosthetic hand that can perform several of the grip functions needed in activities of daily living.

The USA has recently invested a large sum of money in developing whole-arm prostheses. The US Defense Advanced Research Projects Agency (DARPA) has made substantial investments in this field – a project inspired by the many cases of whole-arm amputations among soldiers returning home from wars in Iraq and Afghanistan. When the whole arm is amputated, there are no remaining muscles to generate EMG signals for controlling the prosthesis. To overcome this problem, Todd Kuiken and his colleagues at the Rehabilitation Institute of Chicago described an interesting principle, *targeted reinnervation*, where the remaining proximal parts of the nerves that originally innervated the arm and hand, instead are redirected and connected to muscles in the pectoral region of the frontal aspect of chest [65–71]. Nerve fibres that previously innervated muscles in the forearm and hand now innervate and activate muscles in the pectoral region. The pectoral muscles then act as magnifiers for nerve signals devoted to the forearm and hand, and electrodes placed on top of these muscles allow the EMG signals to control movements in a whole-arm prosthesis through pectoral muscle contractions. With the DEKA arm, which is constructed according to these principles, the user is able to move not only the artificial hand but also the shoulder and elbow. Since the transferred nerves also contain sensory fibres, there is also some sensory reinnervation of coetaneous areas in the pectoral region, which can be used as a substitute for the lost hand sensibility. Hence, when this skin is touched, it provides the amputee with a sense of the missing arm or hand being touched [72]. The principles of targeted reinnervation have opened new vistas for those who are in need of a substitute for a whole amputated arm.

How to Provide a Hand Prosthesis with Sensory Feedback?

A hand without sensibility is 'blind', or at least a hand with very limited function. In the same way, a hand prosthesis without sensibility is never an optimal substitute for a hand. A hand prosthesis without sensory feedback may have a role as a technical aid that is attached to the body, serving to assist the healthy hand, but it will not be perceived as a natural and true part of the body, and the brain will not develop a sense of body ownership of the prosthesis. When there is no sensory function in the hand prosthesis, other senses – vision and hearing – must compensate for the lack of sensory feedback. Accordingly, the prosthesis user must have the prosthetic hand constantly within sight to be able to control its movements. The lack of sensation is one major reason for the restrictive use of hand prostheses among amputees [73, 74].

In clinical trials the so-called rubber hand phenomenon (cf. Chap. 10) can be evoked on hand amputees so that the amputees perceive the hand prosthesis as a feeling part of their own body. This is accomplished by applying synchronous strokes to the residual stump, out of view, and to a rubber hand or hand prosthesis, placed in full view. This elicits an illusion of sensing touch on the artificial hand, rather than on the stump, so a feeling of ownership of the artificial hand develops [75, 76]. However, the phenomenon lasts only as long as the prosthesis user keeps observing the hand prosthesis.

The rubber hand phenomenon can teach us much about how the brain works, but the illusion can only be elicited in experimental situations and is of no practical value in everyday life. Much attention is currently being paid to various principles for achieving sensory feedback and tactile sensory functions in a hand prosthesis [77–85]. Some manufacturers have incorporated 'slip sensors' into prosthetic fingers to detect the vibrations occurring when an item starts to slip out of the grip. This triggers an automatic increase in the grip force [77, 86]. But here we are not dealing with conscious sensibility; it is a closed feedback system in the hand prosthesis itself.

A prerequisite for the hand prosthesis to be perceived as a living and true part of the body is that there is a *conscious* sensory feedback from the prosthesis, preferably combined with useful tactile discriminative functions, so that the user can recognise and define the shape, configuration and texture of the objects that the prosthesis touches [87, 88]. Several different techniques can be employed to provide amputees with sensory feedback [83]. A useful sensory feedback can be achieved in body-powered prosthetic devices where the user can sense the prosthetic state and grip strength through the reaction forces that are transmitted by the control cable and harness on their skin/body. Also, the motor sound from the prosthesis, as well as the socket pressure, may provide useful sensory feedback along with the visual information. Achieving tactile discriminative functions requires sensors in the prosthetic fingers responding to various types of tactile stimuli such as pressure and vibration. In a normal hand, various types of sensory receptors are activated by touch/pressure, vibration, stretching, pain and temperature (cf. Chap. 7). Development of artificial sensors with analogous functions is an important part of developing a complete sensory feedback system in an artificial hand. In normal hand function, mechanoreceptors responding to touch/pressure and vibration are crucial. Sensors in an artificial hand that respond to pressure can be of various types, for instance, piezoelectric membranes or optical systems where increased pressure results in proportionally impaired light conductance or small air-containing silicone pads allowing redistribution of air [83, 84, 89]. Another way to measure force on the fingers is to monitor the tension in artificial flexor tendons when the prosthesis grips an item. Such sensory feedback may be essential for the ability to regulate the grip force.

Sensors for vibratory stimuli may be miniature microphones incorporated in the prosthesis [90] (cf. Chap. 10). When a hand or hand prosthesis moves across a surface, friction causes vibrations. The pattern of vibratory signals is characteristic of the particular texture, and the sound of the touch can be used in hand prostheses to identify surface structures; thus, hearing is substituted for touch. Functional MRI studies have demonstrated that, as a result of audio-tactile interaction, the 'sound of touch' in a trained patient will activate both the hearing area of the brain and the sensory cortex [90, 91].

A major problem is how sensory information from the prosthesis can be made conscious [82, 83]. This requires a man–machine interface between the sensors and the user's nervous system. There are essentially three possible ways to achieve this goal: by conveying sensory information directly to the brain, to afferent nerve trunks or to intact skin containing sensory receptors.

Conveying sensory information directly to the brain via implanted electrodes is a complicated principle, although it has been applied in monkeys by Nicolelis and his research team [92] reporting on the operation of a brain-machine interface that allows signalling of artificial tactile feedback through intracortical microstimulation of the somatosensory cortex. Another way to convey information from an artificial hand to the nervous system would be via electrodes introduced into peripheral nerve trunks in the residual limb [93]. Different types of electrodes can be used, such as cuff electrodes, in which electrode wires are wrapped around the nerve [94], or electrodes placed in nerves longitudinally [36–38, 94]. However, along with causing a number of biological and technical problems, such a system would require stimulation of correct fibre types in the nerve to translate the electrical signals from the hand prosthesis to meaningful modality-specific sensory perceptions. The quality of the sensation perceived by individuals using this system has usually been a foreign feeling resembling paraesthesia, vibration, tapping or flutter on the skin [83].

A more physiological means of conveying sensory information from an artificial hand to the nervous system is to use the skin of the residual limb as man–machine interface. Attempts have been made to create sensory feedback by applying a weak electrical current to the skin; however, the patients perceived this as very unpleasant [83, 95–97]. The sensation was not necessarily confined to the zone under the electrodes, but the elicited sensations could spread if they were placed near nerve bundles. Such electrocutaneous stimulation is not used in clinical practice to achieve sensory feedback from hand prostheses.

The skin of the residual limb contains mechanoreceptors that respond to touch/pressure stimuli and vibratory stimuli (Merkel end organs, Meissner end organs and Pacini corpuscles), and therefore, they can serve as appropriate targets for corresponding stimuli, transferred from tactile receptors in the artificial hand. However, to achieve a natural, physiological sensory perception, the sensory feedback needs to be modality-specific, meaning that the output is felt in the same modality as the sensory input – for example, a touch on the prosthesis should be felt as touch and vibration applied to the prosthesis should be perceived as vibration.

Several studies and clinical set-ups involved conveying sensory signals from the prosthesis to the skin of the amputation stump as vibrotactile stimuli [44, 55, 56, 78, 97–107]. But when applied to the skin, vibrations represent a strange and very unpleasant type of stimuli, and the skin rapidly adapts to the vibrations, making them less effective.

A more physiological and natural way to achieve modality-specific sensory feedback is to transfer touch/pressure stimuli applied to the hand prosthesis to analogous touch/pressure stimuli applied to the skin of the residual forearm. In ongoing research at Lund University in Sweden, this principle is the main track to achieve conscious sensory feedback from a hand prosthesis. The background gives new insights into the interaction between the hand and brain with special focus on the profound reorganisations in the somatosensory cortex that follow hand amputation. Following amputation of a hand, the corresponding representational area in the somatosensory cortex no longer receives any sensory input. Consequently, the adjacent cortical area representing the forearm expands over the 'silent' hand cortical

Fig. 16.3 Phantom hand map distally in the residual forearm of an amputee. In this case the phantom fingers were extended over large areas (From Antfolk et al. [89])

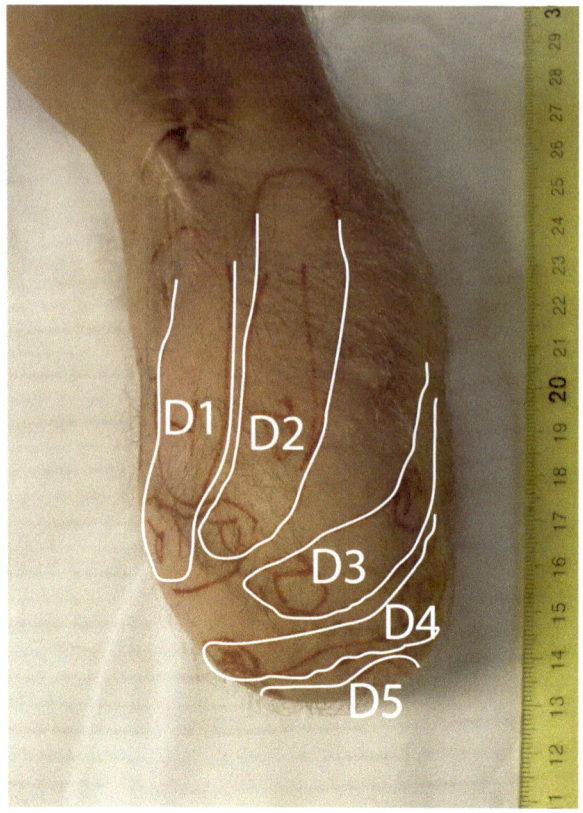

representation. The result is a 'mapping' of the phantom hand in the forearm: on the skin, distally in the residual stump, a 'referred phantom map' of the lost hand occurs [76, 101, 108–111] (Fig. 16.3). When various map areas corresponding to individual fingers are touched, the amputee senses that the corresponding fingers of the missing hand are touched [75, 76, 101, 112], and interestingly, if enough pressure is applied to the various areas of the phantom hand, it activates the corresponding 'correct' areas in the sensory cortex as revealed by fMRI studies [113]. Transferring the information from sensors applied to individual fingers of the prosthetic hand to corresponding finger areas in the referred phantom map in the forearm can provide a physiological and natural sensory feedback, making the prosthesis feel like a part of the body scheme, as proposed by Ehrsson et al. [75]. For this purpose a 'tactile display' can be used, consisting of five actuators (the number of fingers on a hand) to be placed on the phantom hand map on the amputee's residual limb [114]. The actuators may be of various types. One method is to use digital servomotors with small plastic buttons affixed at the end of levers, which press against the underlying skin as the motor rotates [114]. Another possibility is to use small inflatable silicone pads incorporated in the tactile display. The sensory feedback is mediated by air in a closed-loop system connecting silicone pads on the prosthetic hand with pads on the amputation stump. The silicone pads in the tactile display expand when their corresponding sensor bulbs in the prosthesis are touched, evoking a feeling of 'real touch' [89].

The Ideal Hand Prosthesis

The ideal future mind-controlled hand prosthesis should be intuitively able to perform most movements and grip functions that the user performs in activities of daily living. The prosthesis should be well integrated into the user's body and should be perceived as a natural body part. It should function in activities of daily living as well as in the user's occupation and also at leisure. It should be robust and durable.

It should have sensory functions to provide sensory feedback and tactile discriminative functions to enable the user to recognise and identify shapes and textures of items that are touched by the prosthetic hand. It should be easy to remove, but after use it should be easy to reattach to the body either via a fitted sleeve or using screws that are integrated into the bones of the forearm in line with the osseointegration principle developed and described by the Professor Per-Ingvar Brånemark in Gothenburg [115, 116].

The ideal hand prosthesis should be felt as an extension of the brain and should be perceived as a natural part of the body. It should be able to behave like a hand and to feel like a hand. This concept represents one of the biggest medical-technical challenges of our time.

References

1. Sorbye R. Myoelectric prosthetic fitting in young children. Clin Orthop Relat Res. 1980;148:34–40.
2. Billock JN. Prosthetic management of complete hand and arm deficiencies. In: Hunter JM, Mackin EJ, Callahan AD, editors. Rehabilitation of the hand: surgery and therapy. 4th ed. St. Louis: Mosby; 1995. p. 1189–201.
3. Hubbard S. Myoprosthetic management of the upper limb amputee. In: Hunter JM, Mackin EJ, Callahan AD, editors. Rehabilitation of the hand: surgery and therapy. 4th ed. St. Louis: Mosby; 1995. p. 1241–52.
4. Atkins DJ, Heard DCY, Donovan WH. Epidemiologic overview of individuals with upper-limb loss and their reported research priorities. J Prosthet Orthot. 1996;8:2–11.
5. Kyberd PJ, Chappell PH. The Southampton Hand: an intelligent myoelectric prosthesis. J Rehabil Res Dev. 1994;31(4):326–34.
6. Lundborg G. Tomorrow's artificial hand. Scand J Plast Reconstr Surg Hand Surg. 2000; 34(2):97–100.
7. Mercier C, Reilly KT, Vargas CD, Aballea A, Sirigu A. Mapping phantom movement representations in the motor cortex of amputees. Brain. 2006;129(Pt 8):2202–10.
8. Nicolelis MA, Lebedev MA. Principles of neural ensemble physiology underlying the operation of brain-machine interfaces. Nat Rev Neurosci. 2009;10(7):530–40.

9. Kennedy PR, Bakay RA. Restoration of neural output from a paralyzed patient by a direct brain connection. Neuroreport. 1998;9(8):1707–11.

10. Kennedy PR, Bakay RA, Moore MM, Adams K, Goldwaithe J. Direct control of a computer from the human central nervous system. IEEE Trans Rehabil Eng. 2000;8(2):198–202.

11. Birbaumer N, Cohen LG. Brain-computer interfaces: communication and restoration of movement in paralysis. J Physiol. 2007;579(Pt 3):621–36.

12. Birbaumer N, Ghanayim N, Hinterberger T, Iversen I, Kotchoubey B, Kubler A, et al. A spelling device for the paralysed. Nature. 1999;398(6725):297–8.

13. Karim AA, Hinterberger T, Richter J, Mellinger J, Neumann N, Flor H, et al. Neural internet: web surfing with brain potentials for the completely paralyzed. Neurorehabil Neural Repair. 2006;20(4):508–15.

14. Kennedy PR, Kirby MT, Moore MM, King B, Mallory A. Computer control using human intracortical local field potentials. IEEE Trans Neural Syst Rehabil Eng. 2004;12(3):339–44.

15. Nijboer F, Sellers EW, Mellinger J, Jordan MA, Matuz T, Furdea A, et al. A P300-based brain-computer interface for people with amyotrophic lateral sclerosis. Clin Neurophysiol. 2008;119(8):1909–16.

16. Wolpaw JR, Birbaumer N, McFarland DJ, Pfurtscheller G, Vaughan TM. Brain-computer interfaces for communication and control. Clin Neurophysiol. 2002;113(6):767–91.

17. Wolpaw JR, McFarland DJ. Control of a two-dimensional movement signal by a noninvasive brain-computer interface in humans. Proc Natl Acad Sci U S A. 2004;101(51):17849–54.

18. Fischman J. A bionic boost. Nat Geosci. 2010;217:34–53.

19. Schwartz AB, Cui XT, Weber DJ, Moran DW. Brain-controlled interfaces: movement restoration with neural prosthetics. Neuron. 2006;52(1):205–20.

20. Nicolelis MA. Actions from thoughts. Nature. 2001;409(6818):403–7.

21. Nicolelis MA. Brain-machine interfaces to restore motor function and probe neural circuits. Nat Rev Neurosci. 2003;4(5):417–22.

22. Wessberg J, Stambaugh CR, Kralik JD, Beck PD, Laubach M, Chapin JK, et al. Real-time prediction of hand trajectory by ensembles of cortical neurons in primates. Nature. 2000;408(6810):361–5.

23. Chapin JK, Moxon KA, Markowitz RS, Nicolelis MA. Real-time control of a robot arm using simultaneously recorded neurons in the motor cortex. Nat Neurosci. 1999;2(7):664–70.

24. Nicolelis MA, Chapin JK. Controlling robots with the mind. Sci Am. 2002;287(4):46–53.

25. Carmena JM, Lebedev MA, Crist RE, O'Doherty JE, Santucci DM, Dimitrov DF, et al. Learning to control a brain-machine interface for reaching and grasping by primates. PLoS Biol. 2003;1(2):E42.

26. Musallam S, Corneil BD, Greger B, Scherberger H, Andersen RA. Cognitive control signals for neural prosthetics. Science. 2004;305(5681):258–62.

27. Velliste M, Perel S, Spalding MC, Whitford AS, Schwartz AB. Cortical control of a prosthetic arm for self-feeding. Nature. 2008;453(7198):1098–101.

28. Hochberg LR, Serruya MD, Friehs GM, Mukand JA, Saleh M, Caplan AH, et al. Neuronal ensemble control of prosthetic devices by a human with tetraplegia. Nature. 2006;442(7099):164–71.

29. Simeral JD, Kim SP, Black MJ, Donoghue JP, Hochberg LR. Neural control of cursor trajectory and click by a human with tetraplegia 1000 days after implant of an intracortical microelectrode array. J Neural Eng. 2011;8(2):025027.

30. Collinger JL, Wodlinger B, Downey JE, Wang W, Tyler-Kabara EC, Weber DJ, et al. High-performance neuroprosthetic control by an individual with tetraplegia. Lancet. 2013;381(9866):557–64.

31. Eriksson Linsmeier C, Prinz CN, Pettersson LM, Caroff P, Samuelson L, Schouenborg J, et al. Nanowire biocompatibility in the brain–looking for a needle in a 3D stack. Nano Lett. 2009;9(12):4184–90.

32. Thelin J, Jorntell H, Psouni E, Garwicz M, Schouenborg J, Danielsen N, et al. Implant size and fixation mode strongly influence tissue reactions in the CNS. PLoS One. 2011;6(1):e16267.

33. Linsmeier CE, Thelin J, Danielsen N. Can histology solve the riddle of the nonfunctioning electrode? Factors influencing the biocompatibility of brain machine interfaces. Prog Brain Res. 2011;194:181–9.
34. Lind G, Gallentoft L, Danielsen N, Schouenborg J, Pettersson LM. Multiple implants do not aggravate the tissue reaction in rat brain. PLoS One. 2012;7(10):e47509.
35. Stix G. Jacking into the brain. Sci Am. 2008;299(5):56–61.
36. Dhillon GS, Horch KW. Direct neural sensory feedback and control of a prosthetic arm. IEEE Trans Neural Syst Rehabil Eng. 2005;13(4):468–72.
37. Dhillon GS, Lawrence SM, Hutchinson DT, Horch KW. Residual function in peripheral nerve stumps of amputees: implications for neural control of artificial limbs. J Hand Surg Am. 2004;29(4):605–15; discussion 16–8.
38. Rossini PM, Micera S, Benvenuto A, Carpaneto J, Cavallo G, Citi L, et al. Double nerve intraneural interface implant on a human amputee for robotic hand control. Clin Neurophysiol. 2010;121(5):777–83.
39. Warwick K, Gasson M, Hutt B, Goodhew I, Kyberd P, Andrews B, et al. The application of implant technology for cybernetic systems. Arch Neurol. 2003;60(10):1369–73.
40. Almstrom C, Herberts P, Korner L. Experience with Swedish multifunctional prosthetic hands controlled by pattern recognition of multiple myoelectric signals. Int Orthop. 1981;5(1):15–21.
41. Postema K, van der Donk V, van Limbeek J, Rijken RA, Poelma MJ. Prosthesis rejection in children with a unilateral congenital arm defect. Clin Rehabil. 1999;13(3):243–9.
42. Wagner LV, Bagley AM, James MA. Reasons for prosthetic rejection by children with unilateral congenital transverse forearm total deficiency. JPO. 2007;19:51–4.
43. Lewis S, Russold MF, Dietl H, editors. User demands for sensory feedback in upper extremity prostheses. Medical Measurements and Applications Proceedings (MeMeA), 2012 IEEE international symposium. Budapest: Otto Bock Healthcare Products; 2012.
44. Pylatiuk C, Mounier S, Kargov A, Schulz S, Bretthauer G. Progress in the development of a multifunctional hand prosthesis. Conf Proc IEEE Eng Med Biol Soc. 2004;6:4260–3.
45. Sebelius FC, Rosen BN, Lundborg GN. Refined myoelectric control in below-elbow amputees using artificial neural networks and a data glove. J Hand Surg Am. 2005;30(4):780–9.
46. Parker P, Englehart K, Hudgins B. Myoelectric signal processing for control of powered limb prostheses. J Electromyogr Kinesiol. 2006;16(6):541–8.
47. Chan FH, Yang YS, Lam FK, Zhang YT, Parker PA. Fuzzy EMG classification for prosthesis control. IEEE Trans Rehabil Eng. 2000;8(3):305–11.
48. Hudgins B, Parker P, Scott RN. A new strategy for multifunction myoelectric control. IEEE Trans Biomed Eng. 1993;40(1):82–94.
49. Peleg D, Braiman E, Yom-Tov E, Inbar GF. Classification of finger activation for use in a robotic prosthesis arm. IEEE Trans Neural Syst Rehabil Eng. 2002;10(4):290–3.
50. Boostani R, Moradi MH. Evaluation of the forearm EMG signal features for the control of a prosthetic hand. Physiol Meas. 2003;24(2):309–19.
51. Castellini C, Gruppioni E, Davalli A, Sandini G. Fine detection of grasp force and posture by amputees via surface electromyography. J Physiol Paris. 2009;103(3–5):255–62.
52. Tenore FV, Ramos A, Fahmy A, Acharya S, Etienne-Cummings R, Thakor NV. Decoding of individuated finger movements using surface electromyography. IEEE Trans Biomed Eng. 2009;56(5):1427–34.
53. Momen K, Krishnan S, Chau T. Real-time classification of forearm electromyographic signals corresponding to user-selected intentional movements for multifunction prosthesis control. IEEE Trans Neural Syst Rehabil Eng. 2007;15(4):535–42.
54. Zecca M, Micera S, Carrozza MC, Dario P. Control of multifunctional prosthetic hands by processing the electromyographic signal. Crit Rev Biomed Eng. 2002;30(4–6):459–85.
55. Cipriani C, Zaccone F, Micera S, Carrozza MC. On the shared control of an EMG-controlled prosthetic hand: analysis of user-prosthesis interaction. IEEE Trans Robot. 2008;24:170–84.

56. Pons JL, Recon E, Ceres R, Reynaerts D, Saro B, Levin S, et al. The MANUS-HAND dextrous robotics upper limb prosthesis: mechanical and manipulation aspects. Auton Robot. 2004;16:143–63.
57. Nishikawa D, Yu W, Yokoi H, Kakazu Y. On-line learning method for EMG prosthetic hand control. Electron Commun Jap. 2001;84:35–46.
58. Carozza MC, Cappiello G, Micera S, Edin BB, Beccai L, Cipriani C. Design of cybernetic hand for perception and action. Biol Cybern. 2006;95:629–44.
59. Sebelius F, Axelsson M, Danielson N, Schouenborg J, Laurell T. Real-time control of a virtual hand. Technol Disabil. 2005;17:131–41.
60. Carozza MC, Suppo C, Sebastiani F, Massa B, Vecchi F, Lazzarini R, et al. The SPRING Hand: development of a self-adaptive prosthesis for restoring natural grasping. Auton Robot. 2004;16:125–41.
61. Oskoei MA, Hu H. Myoelectric control systems – a survey. Biomed Signal Process Control. 2007;2:275–94.
62. Sebelius F, Eriksson L, Holmberg H, Levinsson A, Lundborg G, Danielsen N, et al. Classification of motor commands using a modified self-organising feature map. Med Eng Phys. 2005;27(5):403–13.
63. Sebelius F, Eriksson L, Balkenius C, Laurell T. Myoelectric control of a computer animated hand: a new concept based on the combined use of a tree-structured artificial neural network and a data glove. J Med Eng Technol. 2006;30(1):2–10.
64. Kanitz GR, Antfolk C, Cipriani C, Sebelius F, Carozza MC. Decoding of individuated finger movements using surface EMG and input optimization applying a genetic algorithm. Conf Proc IEEE Eng Med Biol Soc. 2011;2011:1608–11.
65. Kuiken TA, Dumanian GA, Lipschutz RD, Miller LA, Stubblefield KA. The use of targeted muscle reinnervation for improved myoelectric prosthesis control in a bilateral shoulder disarticulation amputee. Prosthet Orthot Int. 2004;28(3):245–53.
66. Li G, Kuiken TA. EMG pattern recognition control of multifunctional prostheses by transradial amputees. Conf Proc IEEE Eng Med Biol Soc. 2009;2009:6914–7.
67. Marasco PD, Schultz AE, Kuiken TA. Sensory capacity of reinnervated skin after redirection of amputated upper limb nerves to the chest. Brain. 2009;132(Pt 6):1441–8.
68. Tkach D, Huang H, Kuiken TA. Study of stability of time-domain features for electromyographic pattern recognition. J Neuroeng Rehabil. 2010;7:21.
69. Li G, Schultz AE, Kuiken TA. Quantifying pattern recognition-based myoelectric control of multifunctional transradial prostheses. IEEE Trans Neural Syst Rehabil Eng. 2010;18(2):185–92.
70. Kuiken TA. Consideration of nerve-muscle grafts to improve the control of artificial arms. J Technol Disabil. 2003;15:105–11.
71. Schultz AE, Kuiken TA. Neural interfaces for control of upper limb prostheses: the state of the art and future possibilities. PM R. 2011;3(1):55–67.
72. Marasco PD, Kim K, Colgate JE, Peshkin MA, Kuiken TA. Robotic touch shifts perception of embodiment to a prosthesis in targeted reinnervation amputees. Brain. 2011;134(Pt 3):747–58.
73. Biddiss E, Beaton D, Chau T. Consumer design priorities for upper limb prosthetics. Disabil Rehabil Assist Technol. 2007;2(6):346–57.
74. Kyberd PJ, Wartenberg C, Sandsjö L. Survey of upper-extremity prosthesis users in Sweden and the United Kingdom. J Prosthet Orthot. 2007;19:55–62.
75. Ehrsson HH, Rosen B, Stockselius A, Ragno C, Kohler P, Lundborg G. Upper limb amputees can be induced to experience a rubber hand as their own. Brain. 2008;131(Pt 12):3443–52.
76. Rosen B, Ehrsson HH, Antfolk C, Cipriani C, Sebelius F, Lundborg G. Referral of sensation to an advanced humanoid robotic hand prosthesis. Scand J Plast Reconstr Surg Hand Surg. 2009;43(5):260–6.
77. Childress DS. Closed-loop control in prosthetic systems: historical perspective. Ann Biomed Eng. 1980;8(4–6):293–303.
78. Shannon GF. A comparison of alternative means of providing sensory feedback on upper limb prostheses. Med Biol Eng. 1976;14(3):289–94.

79. Nohama P, Lopes AV, Cliquet JA. Electrotactile stimulator for artificial proprioception. Artif Organs. 1995;19(3):225–30.
80. Phillips CA. Sensory feedback control of upper- and lower-extremity motor prostheses. Crit Rev Biomed Eng. 1988;16(2):105–40.
81. Patterson PE, Katz JA. Design and evaluation of a sensory feedback system that provides grasping pressure in a myoelectric hand. J Rehabil Res Dev. 1992;29(1):1–8.
82. Lundborg G, Rosen B. Sensory substitution in prosthetics. Hand Clin. 2001;17(3):481–8, ix–x.
83. Antfolk C, D'Alonzo M, Rosen B, Lundborg G, Sebelius F, Cipriani C. Sensory feedback in upper limb prosthetics. Expert Rev Med Devices. 2013;10(1):45–54.
84. Chappell PH. Making sense of artificial hands. J Med Eng Technol. 2011;35(1):1–18.
85. Riso RR. Strategies for providing upper extremity amputees with tactile and hand position feedback–moving closer to the bionic arm. Technol Health Care. 1999;7(6):401–9.
86. Puchhammer G. The tactile slip sensor: integration of a miniaturized sensory device on an myoelectric hand. Orthopädie-Technik Q. 2000;1:7–12.
87. Lederman SJ, Klatzky RL. Hand movements: a window into haptic object recognition. Cogn Psychol. 1987;19(3):342–68.
88. Klatzky RL, Lederman SJ. Identifying objects from a haptic glance. Percept Psychophys. 1995;57(8):1111–23.
89. Antfolk C, Bjorkman A, Frank SO, Sebelius F, Lundborg G, Rosen B. Sensory feedback from a prosthetic hand based on air-mediated pressure from the hand to the forearm skin. J Rehabil Med. 2012;44(8):702–7.
90. Lundborg G, Rosen B, Lindberg S. Hearing as substitution for sensation: a new principle for artificial sensibility. J Hand Surg Am. 1999;24(2):219–24.
91. Lundborg G, Bjorkman A, Hansson T, Nylander L, Nyman T, Rosen B. Artificial sensibility of the hand based on cortical audiotactile interaction: a study using functional magnetic resonance imaging. Scand J Plast Reconstr Surg Hand Surg. 2005;39(6):370–2.
92. O'Doherty JE, Lebedev MA, Ifft PJ, Zhuang KZ, Shokur S, Bleuler H, et al. Active tactile exploration using a brain-machine-brain interface. Nature. 2011;479(7372):228–31.
93. Micera S, Carpaneto J, Raspopovic S. Control of hand prostheses using peripheral information. IEEE Rev Biomed Eng. 2010;3:48–68.
94. Horch K, Meek S, Taylor TG, Hutchinson DT. Object discrimination with an artificial hand using electrical stimulation of peripheral tactile and proprioceptive pathways with intrafascicular electrodes. IEEE Trans Neural Syst Rehabil Eng. 2011;19(5):483–9.
95. Lundborg G, Rosen B, Lindstrom K, Lindberg S. Artificial sensibility based on the use of piezoresistive sensors. Preliminary observations. J Hand Surg Br. 1998;23(5):620–6.
96. Hernandez Arieta A, Yokoi H, Arai T, Yu W. Study on the effects of electrical stimulation on the pattern recognition for an EMG prosthetic application. Conf Proc IEEE Eng Med Biol Soc. 2005;7:6919–22.
97. Kaczmarek KA, Webster JG, Bach-y-Rita P, Tompkins WJ. Electrotactile and vibrotactile displays for sensory substitution systems. IEEE Trans Biomed Eng. 1991;38(1):1–16.
98. Cipriani C, D'Alonzo M, Carrozza MC. A miniature vibrotactile sensory substitution device for multifingered hand prosthetics. IEEE Trans Biomed Eng. 2012;59(2):400–8.
99. Jones LA, Sarter NB. Tactile displays: guidance for their design and application. Hum Factors. 2008;50(1):90–111.
100. Wilska A. On the vibrational sensitivity in different regions of the body surface. Acta Physiol Scand. 1954;31(2–3):284–9.
101. Antfolk C, D'Alonzo M, Controzzi M, Lundborg G, Rosen B, Sebelius F, et al. Artificial redirection of sensation from prosthetic fingers to the phantom hand map on transradial amputees: vibrotactile versus mechanotactile sensory feedback. IEEE Trans Neural Syst Rehabil Eng. 2013;21(1):112–20.
102. Chatterjee K, Guo Z, Vogler EA, Siedlecki CA. Contributions of contact activation pathways of coagulation factor XII in plasma. J Biomed Mater Res A. 2009;90(1):27–34.

103. Stepp CE, Matsuoka Y. Relative to direct haptic feedback, remote vibrotactile feedback improves but slows object manipulation. Conf Proc IEEE Eng Med Biol Soc. 2010; 2010:2089–92.

104. Bach-y-Rita P, Collins CC, editors. Sensory substitution and limb prosthesis. In: Proceedings of the third international symposium on advances in external control of human extremities; 1970.

105. Mann RW, Reimers SD. Kinesthetic sensing for the EMG controlled Boston arm. IEEE Trans Man Mach Syst. 1970;MM11:110–5.

106. Pons JL, Ceres R, Rocon E. Objectives and technological approach to the development of the multifunctional MANUS upper limb prosthesis. Robotica. 2005;23:301–10.

107. Pylatiuk C, Kargov A, Schulz S. Design and evaluation for myoelectric prosthetic hands. J Prosthet Orthot. 2006;18:57–61.

108. Ramachandran VS, Blakeslee S. Phantoms in the brain: human nature and the architecture of the mind. London: Fourth Estate; 1999.

109. Borsook D, Becerra L, Fishman S, Edwards A, Jennings CL, Stojanovic M, et al. Acute plasticity in the human somatosensory cortex following amputation. Neuroreport. 1998; 9(6):1013–7.

110. Ramachandran VS. Behavioral and magnetoencephalographic correlates of plasticity in the adult human brain. Proc Natl Acad Sci U S A. 1993;90(22):10413–20.

111. Schmalzl L, Thomke E, Ragno C, Nilseryd M, Stockselius A, Ehrsson HH. "Pulling telescoped phantoms out of the stump": manipulating the perceived position of phantom limbs using a full-body illusion. Front Hum Neurosci. 2011;5:121.

112. Schmalzl L, Ragno C, Ehrsson HH. An alternative to traditional mirror therapy: illusory touch can reduce phantom pain when illusory movement does not. Clin J Pain. 2013 Feb 26.

113. Bjorkman A, Weibull A, Olsrud J, Ehrsson HH, Rosen B, Bjorkman-Burtscher IM. Phantom digit somatotopy: a functional magnetic resonance imaging study in forearm amputees. Eur J Neurosci. 2012;36(1):2098–106.

114. Antfolk C, Balkenius C, Rosen B, Lundborg G, Sebelius F. SmartHand tactile display: a new concept for providing sensory feedback in hand prostheses. Scand J Plast Reconstr Surg Hand Surg. 2010;44(1):50–3.

115. Brånemark PI. The osseointegration book. From calvarium to calcaneus. Berlin: Quintessence; 2005.

116. Lundborg G, Branemark PI, Rosen B. Osseointegrated thumb prostheses: a concept for fixation of digit prosthetic devices. J Hand Surg Am. 1996;21(2):216–21.

Index

G. Lundborg, *The Hand and the Brain*,
DOI 10.1007/978-1-4471-5334-4, © Springer-Verlag London 2014